BASICS OF UNMANNED AERIAL VEHICLES

TIME TO START WORKING ON DRONE TECHNOLOGY

GARVIT PANDYA

INDIA • SINGAPORE • MALAYSIA

Notion Press

No.8, 3rd Cross Street
CIT Colony, Mylapore
Chennai, Tamil Nadu – 600004

First Published by Notion Press 2021
Copyright © Garvit Pandya 2021
All Rights Reserved.

ISBN 978-1-63745-386-5

This book has been published with all efforts taken to make the material error-free after the consent of the author. However, the author and the publisher do not assume and hereby disclaim any liability to any party for any loss, damage, or disruption caused by errors or omissions, whether such errors or omissions result from negligence, accident, or any other cause.

While every effort has been made to avoid any mistake or omission, this publication is being sold on the condition and understanding that neither the author nor the publishers or printers would be liable in any manner to any person by reason of any mistake or omission in this publication or for any action taken or omitted to be taken or advice rendered or accepted on the basis of this work. For any defect in printing or binding the publishers will be liable only to replace the defective copy by another copy of this work then available.

DEDICATION

With deepest gratitude and respect,
I dedicate this book to my technical guru

Late Yogendra Jahagirdar Sir
(Jahagirdar Aero Products, Ahmednagar)

who has been a constant source of knowledge and inspiration
in my development in the field of Unmanned Aerial Vehicles and to

Late Dilip Mehta Sir
(Mehta's Institute of English, Rajkot)

A guide who helped me in developing my writing skills
as well as gave me many important life lessons

CONTENTS

Acknowledgement   7

Preface   9

Abbreviations   11

1. Unmanned Systems/Vehicles   15
2. Introduction to UAVs and Drones   21
3. UAV Categories/Classification   27
4. Types of Unmanned Aerial Vehicles   41
5. Multirotor UAV Systems   69
6. On-Board Electrical Components   91
7. Uav Payloads   113
8. Remote Controller and Communications   123
9. Autopilot System and Sensors   133
10. Ground Control Station System   143
11. Advantages of UAVs   157
12. Application of UAVs   161

Contents

Appendix A	*165*
Appendix B	*167*
Appendix C	*173*
Appendix D	*175*
Appendix E	*179*
Appendix F	*183*
Reference Links	*187*

ACKNOWLEDGEMENT

I would like to gratefully acknowledge various people who were a part of my life during the preparation of this book. I am thankful to the companies who permitted me to use their images for explanation purposes and to all those who put images on the public domain to be used without permission for mutual benefits. I am thankful to all the friends, colleagues and other connections that were directly or indirectly, knowingly or unknowingly, involved in my writing process. May God richly bless you all!

My special thanks to my parents, who always provided their support while I was writing this book, especially my father for his support in reviewing the draft of this book.

PREFACE

In the 21st century, the world is evolving, and technology is advancing at a very fast pace. Skills are now getting much more important than before. As the world moves ahead towards the future of technology, AI-based Flying Robots are quite ahead with speedy research going on everywhere in the field. Thus, it is of much importance for all Drone enthusiasts and technology lovers out there to know more about the Drones. This is equally important for the upcoming engineers, as the future technologies are dependent on multidisciplinary skills.

To definitely move ahead as per the scenario, change in the technology is an important decision and a source of generation of new energy. But to know about technologies in advance, one will need to gain the required knowledge now, with a proper vision. There are upcoming engineers and the now growing children who play with nano drones or want to play with them, but they are not getting them as they may have some age restrictions. There are above age enthusiasts, maybe someone with the age of 40-45 or even higher, who may not go out to work on any of the future technologies but are quite interested in getting knowledge of the technologies, having vision and imagining about something that is going to spread across the world during their old age.

For such a straight-forward purpose of sharing the basic knowledge, information and what is going on in the field of drones, the book is designed in such a way that it comes out to be an introductory version about all the main types of Unmanned Aerial Vehicles one needs to know. It may help in advancing their career as well as to work practically on certain drone-based projects. The book does not include any calculation for the designing and development of the drones.

The book deals with information about types of drones with an explanation about their sub-parts. The sub-systems can be electronic systems, engines or frames that will be required for the design and development of a drone. The electronics and mechanical sub-systems can also be further divided into something that is useful for flying drones as well as controlling it. The book covers all the aspects important for a basic drone and all the electronics that are involved in the development of Multirotor types of drone. The information includes the operation of a drone, controlling drones, autonomous flying, etc.

The basic knowledge does not remain limited to this book and I, being a specialised UAV Engineer, encourage that all the readers and enthusiasts of any age should start working on this technology. If there are issues with the costing of drones, one can take help from their school and college for project sponsorship. The wholesome meaning of this encouragement is just that it is a mine of knowledge which is getting more and more innovative with the everyday passes. If one starts working practically today, he will be nearer to the future innovation than someone who starts tomorrow.

"The future is near and the sky is the limit."

ABBREVIATIONS

AOA — Angle of Attack

ADSB — Asynchronous Data Surveillance Board

AGM — Air to Ground Missile

ASM — Air to Surface Missile

ASV — Autonomous Surface Vehicle

ATC — Air Traffic Control

AUW — All-up weight

BLDC — Brushless DC Motor

BVLOS — Beyond Visual Line of Sight

CFRP — Carbon Fibre Reinforced Plastics

CG — Centre of Gravity

CW — Clockwise

CCW — Counter Clockwise

DARPA — Defence Advanced Research Projects Agency

DGCA — Directorate General of Civil Aviation

EO/IR — Electro-Optical Infra-Red

ESC — Electronic Speed Controller

Abbreviations

FCS	–	Flight Control System
FOV	–	Field of View
GCS	–	Ground Control Station/System
GFRP	–	Glass Fibre Reinforced Plastics
GNSS	–	Global Navigation Satellite System
GPS	–	Ground Positioning System
HALE	–	High Altitude Long Endurance
HUD	–	Head-up Display
IC	–	Internal Combustion
IMU	–	Inertial Measurement Unit
INS	–	Inertial Navigation System
LiPo	–	Lithium Polymer
Li-Ion	–	Lithium Ion
LOS	–	Line of Sight
MALE	–	Medium Altitude Long Endurance
MAV	–	Micro Aerial Vehicle
mAH	–	mili Ampere Hours
MEMS	–	Micro Electro Mechanical System
MOCA	–	Ministry of Civil Aviation
MTOW	–	Maximum Take-off Weight
NAV	–	Nano Aerial Vehicle
NDVI	–	Normalised Difference Vegetation Index
NiMH	–	Nickel Metal Hydroxide

NiCd	–	Nickel Cadmium
NPNT	–	No Permission No Take-off
PDB	–	Power Distribution Board
PWM	–	Pulse Width Modulation
RC	–	Remote Controller
RF	–	Radio Frequency
ROUV	–	Remotely Operated Underwater Vehicle
RPAS	–	Remotely Piloted Aircraft System
RPM	–	Revolution Per Minute
RTF	–	Ready to Fly
RTH	–	Return to Home
RTK	–	Real Time Kinematics
RTL	–	Return to Launch/Land
STOL	–	Short Take-off and Landing
UAS	–	Unmanned Aircraft System
UAV	–	Unmanned Aerial Vehicle
UGV	–	Unmanned Ground Vehicle
USAF	–	United States Air Force
USV	–	Unmanned Surface Vehicle
UUV	–	Unmanned Underwater Vehicle
VTOL	–	Vertical Take-off and Landing

CHAPTER 1
Unmanned Systems/Vehicles

1.1 Introduction

Unmanned vehicles are basically different kinds of systems that carry out certain tasks pre-decided and pre-assigned to be performed, which mainly covers the applications from commercial inspections to border security, from bomb detection to targeting an enemy vehicle. Unmanned Systems are important for any nation, and their versatility in terms of usage is tremendous to an extent of doing almost anything without any requirement of a physical human presence in any of the unmanned vehicles. There are certain tasks that are necessary to be carried out in air, on the land, on the water, under the water, hazardous environment, between the animals, above certain altitudes, under certain depths of the sea, etc., but they are impossible for human beings to be carried out due to inaccessibility. For accomplishing such tasks, we are now in an era of unmanned technologies which covers all the existing and upcoming unmanned systems. Let us see what different unmanned technologies are heading our day-to-day life towards with commercial or military applications.

1.2 Unmanned Aerial Vehicles

UAV stands for **Unmanned Aerial Vehicles**. Looking at the words separately, it defines their meaning as it should, and it makes sense collectively too. The word **Unmanned** describes that there is no human being on-board; meaning there is no person going to sit inside or on the vehicle, neither a passenger nor a pilot to fly the vehicle. The word **Aerial** describes that it is related to the air and it is meant for flying. The word **Vehicle** stands for the mode of transportation; meaning it has something inside, which can give it a motion. When the abbreviations U, A, and V are combined, it makes UAV. It makes sense; collectively, it refers to a type of vehicle which can fly without any physical presence of the human being/pilot.

Image 1.1   RQ11 Raven UAV[i] (Credit: Arctic Warrior)

Chapter 2 onwards, the book is all about unmanned aerial vehicles. There you will get to know more about drones.

1.3 Unmanned Ground Vehicles

UGV is a ground vehicle operating on the ground without any human driver's presence. This type of vehicle can be operated remotely from the ground control station. The sensors attached to the vehicle will play a vital role in the avoidance of obstacles. UGVs can be used for the purpose of detecting and disabling explosives, surveillance and transfer of logistics indoor and outdoor. One of the most successful operations and applications of a Rover is Curiosity Rover that was sent by NASA, landed on Mars in September 2011. Even after nine years, it is working well.

Image 1.2 Mars Curiosity Rover[ii] (Credit: NASA/JPL-Caltech/MSSS)

1.4 Unmanned Surface Vehicles

An unmanned boat floats and soars on the water surface with the help of radio controls, without any physical captain piloting it. These vehicles are used in oceanography, marine ecosystem survey, fisheries survey and weather data collection. These are also known as sail drones.

It is possible to achieve longer endurance with a USV compared to a manned boat/surface vehicle. Such boats are designed to collect plastic and other waste material from a lake, a river or a sea. Also, micro-level border patrolling is possible with automatic surface boats with the high-resolution camera as night vision systems.

Image 1.3 Autonomous Robotic Systems Operating in the Maritime Domain[iii] (Credit: Office of Naval Research)

1.5 Unmanned Underwater Vehicles

UUVs are also known as ROUVs (Remotely Operated Underwater Vehicles) and AUVs (Autonomous Underwater Vehicles). These vehicles are operated from the surface by a pilot using a remote controller. It is mostly used for ocean research, mine detection, sea floor mapping, weather and temperature data gatherings, check pipelines for any damage, etc.

Unmanned Systems/Vehicles

Image 1.4 Unmanned Underwater Vehicle[iv] (Credit: U.S. Naval Forces Central Command/U.S. Fifth Fleet)

The above unmanned technologies are growing with innovation in their application and advancing faster. In this book, we will talk about Unmanned Aerial Vehicles. Let's move ahead with the introduction and history of Unmanned Aerial Vehicles.

Chapter Insight

Unmanned Vehicle is not limited to air. It can be anything that flies, swims, rovers or performs anything in hybrid mode, as per the environment. More or less, there can be any type of vehicle made for unmanned operation with the help of certain research and development for the fulfilment of some undoable tasks by human-based applications.

CHAPTER 2
Introduction to UAVs and Drones

2.1 Introduction to UAVs

Unmanned Aerial Vehicles (UAV) are also known as Drones. UAVs are pilotless and non-crewed aircraft that are capable of flying as per the guidance received from remote controls or ground controlled stations as well as preloaded desired areas of interest. Sometimes, it can be both. These types of aircraft are known differently as Unmanned Aerial System (UAS), Remotely Piloted Vehicle (RPV), Remotely Piloted Aircraft System (RPAS) and many other names. In fact, the name drone is quite well known among society.

2.2 History of UAVs

Drones may have encountered public mind from the year 2000 or later, but it has got a long history as a stepping stone towards technological advancement and now in the main course of Future Technology.

In 1783, the paper manufacturer brothers in France, Joseph-Michel and Jacques-Etienne Montgolfier, who invented the process of manufacturing transparent paper, became the architects of the first known unmanned balloon flight, also known as Globe Aérostatique. Joseph-Michel hit the idea of flying vehicles with the observation of laundry drying over a fire that formed air pockets of hot air, rising upward and thereby lifting the fabric.

1849 – Unmanned Balloons

It was in August 1849 when the Austrians attacked Venice by sending around 200 pilotless balloons loaded with bombs controlled by timed fuses. The target was not much achieved as only a few bombs exploded as planned while other balloons missed target due to the wind.

Image 2.1 Unmanned Balloons attacking Venice[v]

1916 – Ruston Proctor Aerial Target

In 1916, the Ruston Proctor Aerial Target UAV was invented. It was the first pilotless winged aircraft and can be regarded as the beginning of drones. In the same year, Hewitt-Sperry's Automatic Aeroplane led us ahead towards UAV research and development.[vi]

Image 2.2   Hewitt-Sperry's Automatic Airplane[vii]

1982 – IAI Scout UAV

Israel's IAI Scout was developed in 1970. It was first operated in combat missions by the South African Defence Force against Angola. The drone was able to carry 38 Kg of payload.[viii]

1995 – Predator RQ - 1

Predator UAV was developed by General Atomics for the purpose of reconnaissance for United States Air Force (USAF). This comes under the category of MALE Drone with an endurance of around 24 hours, it is used by many countries for border surveillance. The Predator's updated models are also capable of carrying missiles.

Image 2.3   MQ-1/RQ-1 Predator[ix]

2000-2005 – Multirotor Systems

Multirotor systems are the currently trending drones that are used mostly in commercial applications. It started coming into the market around the year 2000. It has been researched and developed at a very fast pace.

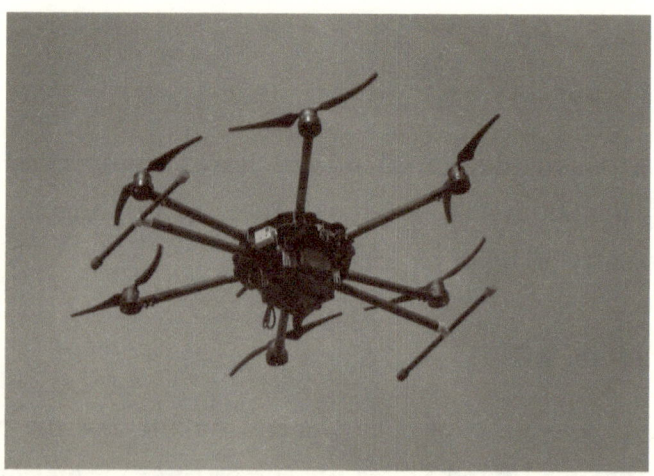

Image 2.4   Hexacopter Drone[x] (Credit: Viktar Masalovich)

Currently Trending Drones and Applications

Drones and applications that are under research and development as well as on the verge of starting trials include hybrid VTOL drones drones for delivery, etc. Additionally, the research is going ahead at a much faster pace for miniaturisation of drones with a focus on application-based nano drones.

2.3 Why we need it?

Drones mainly came into existence from the First World War and the Second World War. During both periods, the main aim was to target the enemy with minimal loss. Unmanned Aircraft were used in a manner so they can fire missiles and drop bombs on enemies at any time, under any climatic condition, without the loss of human beings. This usage of drone later defined into the vehicle that can be used wherever it is difficult for a human being to reach out – like a valley, in between a river, between the mountains during landslides and other natural calamities.

With the technology advancement, the miniaturisation decreased the size of the drone from a large aircraft to something such as quad copter or a Multirotor system. With the changes and evolution in the design of unmanned aircraft (including Multirotor systems), the new applications kept on advancing the society.

Chapter Insight

Drones and Unmanned Aerial Vehicles have been in the syllabus of evolution for quite a long time now. However, with technological advancement for other systems like computers, programming languages, electronic chips, etc., it is now possible to advance at a really faster pace. This effect can be seen in the history that certain things took time for around 100 years to get developed as per the need, whereas these days, updates and upgrades come in a very short period of time, i.e., maybe in a year, on a fair mark.

CHAPTER 3
UAV Categories/Classification

The categories of Unmanned Aerial Vehicles are defined on the basis of its dimensions, maximum take-off weight (payload carrying capacity), maximum flying altitude, range of operation and many other parameters. There is no common standard defined for the classification of UAVs. This means that the categories for all the countries are more or less the same. However, the parameter limitation of any category for two different countries may not be identical. For example, a small drone, as per the U.S. Department of Defence, should have a maximum take-off weight of 9 kg[xi], whereas a small drone in India can weigh anywhere between 2 and 25 kg. In order to know the category of drones in your country, please consider visiting the Government website for information on the same.

3.1 UAV Classification (International Standard)

1. **Small UAVs**

   a) **Nano Drones**

   As the name suggests, nano drones are a kind of mosquito or bee in size. With the small size, weight too decreases. It is necessary that their sub-parts viz. camera, propulsion system and flight controllers are made small enough that they collectively be considered under nano category when fully developed. This category of drone is very small in size but effective for a short range of surveillance.

b) Micro Drones (Micro UAVs or MAVs)

MAVs are no more than 150mm in size and are capable of flying only inside any enclosed premises.

Note: MAVs should not be confused with the micro drones categorised by DGCA in India.

c) Mini Drones

The mini drone is categorised with conditions of MTOW of 20 kg maximum. This type includes Fixed-wings, hexacopters, octocopters and helicopters.

2. Medium Range UAVs

Medium-range UAVs, also known as tactical UAVs, have a range of order between 100 and 300 km. These air vehicles are smaller and operate within simpler systems.

3. MALE UAVs

Medium Altitude Long Endurance Drones (MALE UAVs) fly for around 24 hours at an altitude of up to 15,000 metres. The range for a MALE drone may exceed around 500 km for operation from a ground control station.

4. HALE UAVs

High Altitude Long Endurance Drones (HALE UAVs) fly above 15,000 metres of altitude for more than 24 hours. These drones are used for extremely long range for surveillance purposes.

This also includes drone operations in across the border and cross-border issues.

3.2 UAV Classification (Indian Standard)

Indian Categories for Drone

As per Indian standards, the following categories have been decided by DGCA (Directorate General of Civil Aviation).[xii]

Nano Drones: They weigh less than or equal to 250 grams.

Micro Drones: Their weight is between 250 grams and 2 kg. They can be of any size with 2 kg of MTOW. MAVs can be tricopters, quadcopters, ornithopters and Fixed-wings.

Small Drones: Their weight is between 2 kg and 25 kg. They can be of any size with 25 kg of MTOW. These can be quadcopters, hexacopters, octocopters, helicopters and Fixed-wings.

Medium Drones: They weigh between 25 kg and 150 kg. They can be of any size with 150 kg of MTOW. These can be octocopters, helicopters and Fixed-wings.

Large Drones: They weigh more than 150 kg. They can be of any size with 150 kg of MTOW. These can be octocopters, helicopters and Fixed-wings.

All the categories explained below are as per the Indian standards. The categories for UAVs are different in various countries from India.

3.3 Nano Aerial Vehicles (NAVs)

Nano drones are advanced systems that carry the application of drones to a new level mainly for defence purposes. These applications mainly include surveillance in secret missions inside forests and border securities. These drones are quite small in size and fast which makes it quite difficult to identify them visually and even by radars.

Examples

1. Black Hornet

Prox Dynamics designed and developed a nano helicopter named "Black Hornet" with the application of looking around the corners and walls, forest areas and other obstacles to identify enemy positions by transmitting live video streams to the ground controller in the range of 1.5 km.

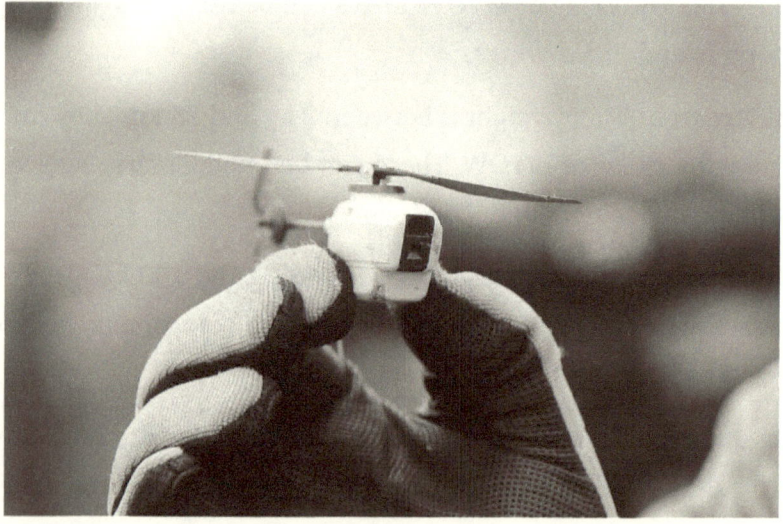

Image 3.1 A Black Hornet Nano UAV from Prox Dynamics (FLIR Systems) (Credit: Richard Watt/MOD)[xiii]

This drone sizes around 10cm x 2.5cm and weighs 16 grams. It's small enough to fit in a hand to launch and land. It is equipped with three cameras, one facing forward, the other facing downward, and the third one facing at a 45° angle. It can fly at a speed of 16 to 18 kmph for 25 minutes.

The drone is currently serving nations like Norway, USA, Australia, Spain, Germany, Dutch Army, France, etc.[xiv]

(Prox Dynamics AS is now owned by FLIR Systems.)

2. Nano Hummingbird

AeroVironment, in the USA, developed a Nano Hummingbird that looks and flies like a humming bird, for Defence Advanced Research Projects Agency (DARPA), in 2011.[xv] This comes under the category of flapping wing vehicles or flapping wing drones. The wing flaps like a bird for propulsion and control of the vehicle.

Image 3.2 Nano Hummingbird[xvi] (Credit: DARPA)

The prototype of AeroVironment Hummingbird has a wingspan of 16.5cms, weighs around 18 to 20 grams and flies for around 10 minutes. This drone can be used for reconnaissance and surveillance in the battlefields. It can also enter buildings to observe the surroundings.

Such flapping wing vehicles are considered under the category of ornithopter.

3.4 Micro Aerial Vehicles

This type of miniaturised drone is mostly used for consumer-based applications like land mapping, for various surveys and for several inspection-purposes being carried out with the help of different types of sensor available.

These drones range in size from around 400mm to 600mm, in the case of Multirotor system. A Fixed-wing drone with a wingspan of around 1 to 1.5 metres may also come under the 2 kg category if the right electronics are selected in order to adjust in the weight compensation. Fixed-wing glider models may have a longer wingspan but come under the 2 kg category.

Examples

1. Typhoon H

An advanced aerial photography and videography drone designed by Yuneec is Typhoon H. It is a hexacopter that weighs less than 2 kg and flies for approximately 25 minutes.

Image 3.3 Yuneec Typhoon H[xvii] (Credit: Webagentur Meerbusch)

Typhoon H flies at the speed of 70 km/hr and has transmitter communication using Wi-Fi with a range of 1.6 km for video transmission.

2. DATAhawk Lite

QuestUAV, a mature UK Aerospace Drone company, has come up with a flying wing/Fixed-wing model, DATAhawk Lite. It is a survey drone under the 2 kg category with a take-off weight capacity of 1995 grams. The wingspan of this drone is 1164mm and flies for around 45 minutes with the range of flying to 20 km at 70 km per hour cruising speed.

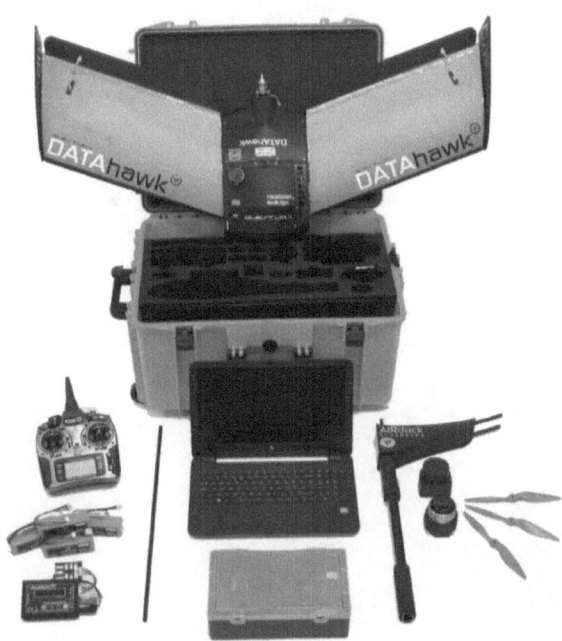

Image 3.4 The DATAhawk Drone[xviii]

DATAhawk lite can carry a Sony QX1 series camera that can be used for mapping. This may be kept ungimballed for a lite category, but with an upgraded version that weighs around 2.2 kg, a gimbal can be useful for various survey and inspection missions.

3.5 Small Aerial Vehicles

Drones which have MTOW of more than 2 kg and less than 25 kg are considered under the small drone category. This category includes drones that can be used for applications like surveillance, film photography, spraying pesticides in agriculture, delivery, etc.

Examples

1. ALTI UAS Transition

It is a multipurpose hybrid Fixed-wing drone that can be used for surveillance, mapping and photogrammetry depending on the sensor attached to the drone. The wingspan of this drone is 3 metres and its MTOW is 18 kg. The MTOW includes frame weight, RTF, fuel and batteries and payloads. This drone is a hybrid model with four electric motors for vertical take-off and landing and one gasoline engine for forward thrust. It has a payload capacity of 1.5 kg, a flying speed of 72 km/hr, a range of 100 km for tracking of the drone and an endurance of around 12 hours.

Image 3.5 ALTI UAS Transition[xix] (Credit: ALTI UAS)

This drone can be used for day and night surveillance with an EO/IR dual sensor with gimbal as well as it can be used for mapping and photogrammetry with a high resolution camera fixed mounted on the drone.

2. Sprayer Drone

Sprayer drone is used for spraying pesticides and water on agricultural farms. Agriculture spraying drones are designed in such a way that they can carry a desired amount of liquid. Most sprayer drones are of size according to the 5 litres, 10 litres and 16 litres of liquid carrying capacity. Below shown drone can carry 10 litres of liquid for spraying on crops in agricultural fields. The maximum weight this drone can carry at the time of take-off is around 25 kg, which includes 10 litres of spraying, batteries and all that it needs to fly efficiently.

Image 3.6 Sprayer Drone[xx] (Credit: Herney Gomez)

With all these loaded over it, such drones fly for approximately 12 to 15 minutes on different paths as assigned and can spray through the terrains with terrain following mode. All these missions are carried out autonomously along with the intelligent and autonomous spray control technology and empty tank warning like a low battery warning notification.

3.6 Medium Aerial Vehicles

With the capacity to carry weight above 25 kg, this category's drones are quite big in dimensions and can be used for anything from cargo delivery to fire fighting. These drones are heavy, of course, because of payloads. However, with the increase in size of the payload, the frame size of the vehicle as well as its weight also increases, along with the high-end heavy propulsion system and power source.

Example

1. **Vapor55**

    Vapor55 is a fully electric helicopter drone with a 2.2 metres rotor span and a copter length of 1.9 metres. The helicopter drone can fly for approximately 1 hour at a cruising speed of 50 km/hr with a total maximum take-off weight of 25 kg, which includes up to 4.5 kg payload weight.

    Image 3.7 Vapor55 Helicopter Drone[xxi] (Credit: Office of Naval Research)

Vapor55 is the same drone that can be used for multiple applications. The applications are based on sensor, and the drone is provided with the space for various sensors carrying capacity and integration facilities. The payload options provided in Vapor55 by AeroVironment are LIDAR sensor, EO/IR camera sensor, hyperspectral sensor, drop mechanism (triggered through flight controller), etc.

3.7 Large Aerial Vehicles

The drones under this category weigh more than 150 kg. These drones are mostly military drones. Military drones will include some as surveillance drones and some as unmanned combat/fighter aircraft that carry guided missiles/bombs and other warfare payloads.

Example

1. MQ-9 Repear/Predator-B

Predator-B, used by the US Air Force (USAF), is designed and developed by General Atomics Aeronautical Systems, first in 2001 and is being developed for its primary use in USAF.

Image 3.8 MQ-9 Reaper RPAS[xxii] (Credit: "Crown copyright 2011" – MOD UK)

Specifications

Weight – 4,800 kg

Length – 36.2 ft

Wingspan – 64 ft

Payload – 360 kg

Ceiling – 45,000 ft

Radius – 400 nm

Endurance – More than 24 hours

Reaper drones are larger and more capable. It has a turboprop engine version of the Air Force MQ-1 Predator, developed as high altitude endurance UAV for science payloads. It can carry up to four Hellfire missiles.[xxiii]

Hellfire missile, also known as AGM-114 Hellfire, is an Air-to-Surface Missile (ASM) first developed for anti-armour use, but later models were developed for precision strikes against different types of targets. The Hellfire missile is the primary 100-pound (45 kg) class air-to-ground precision weapon for the armed forces of the United States and many other nations.

UAV Categories/Classification

Image 3.9   Hellfire ASM Missile[xxiv] (Credit: Hoyasmeg)

Overall, the above drones do exist and they do fulfil desired applications. What are these drones? What are all the common subparts of a drone? How does a Multirotor system fly? All these questions will be answered in the upcoming chapters.

Chapter Insight

So, basically, it can be seen that there are no single particular types of UAVs available. We have all the types as per the classification and that ranges from a few hundred grams to few thousand kilos, mainly dependent on the role they need to perform.

CHAPTER 4
Types of Unmanned Aerial Vehicles

Unmanned Aerial Vehicles are designed with reference to certain aerodynamic techniques and parameters based on category requirements, applications, usability, manoeuvrability requirements, payload configurations, mission requirements, etc. With time, technology advanced in order to meet the requirements and fulfil the purposes. There are certain limitations to Fixed-wing drones. Similarly, there are limitations to Multirotor systems and other vehicles. Let us take a look at the different types of unmanned aerial vehicles available for applications-based operation.

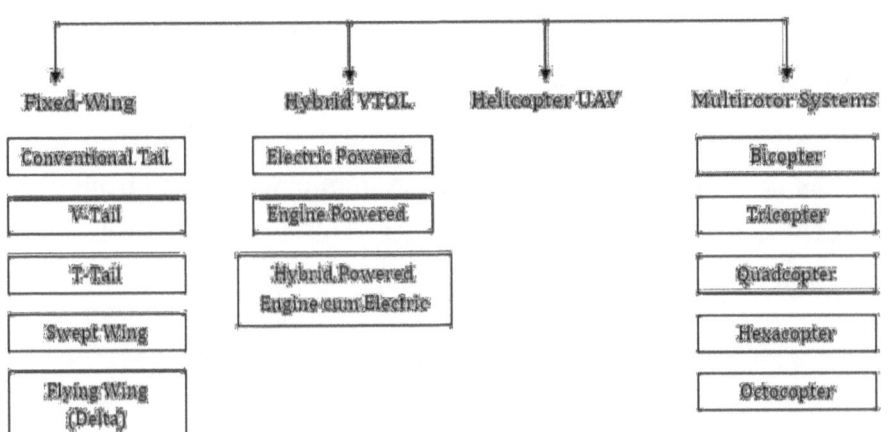

Classification of Unmanned Aerial Vehicles

- Fixed-Wing
  - Conventional Tail
  - V-Tail
  - T-Tail
  - Swept Wing
  - Flying Wing (Delta)
- Hybrid VTOL
  - Electric Powered
  - Engine Powered
  - Hybrid Powered Engine-cum-Electric
- Helicopter UAV
- Multirotor Systems
  - Bicopter
  - Tricopter
  - Quadcopter
  - Hexacopter
  - Octocopter

4.1 Fixed-Wing Vehicles

1. Introduction

Fixed-wing drones have the shape of an aircraft or a scaled model of a conventional aircraft which can be operated with no human pilot onboard. They can be controlled using remote control and ground control stations with the capability of being operated autonomously. They are also named as Fixed-wing UAV, Fixed-wing UAS (Unmanned Aerial System) and Fixed-wing RPAS (Remotely Piloted Aerial System).

2. Scale of Fixed-Wing UAV Based on Operation

Drone dimensions are identified based on application purposes. The application purpose plays a vital role in the calculation of drone sizing, directly or indirectly, as it gives the clear idea about the payload to be used. With the weight coming in to the calculation, all other calculations can be put up accordingly in order to design a scaled plane. Many a time, it is possible that your application can be fulfilled by applying your payload into the existing design.

With the size being defined and the approximate weight coming over in the calculation, it opens up the door for the selection of other parameters in the design and in practical terms of implementation.

Additionally, what type of propulsion system should be used and the placement position of it should be identified at this stage. Fixed-wing drone can have a propulsion system placed in the front of the aircraft, rear and also on the wings. This totally depends on the design of the aircraft.

Types of Unmanned Aerial Vehicles

Image 4.1 Front Engine Fixed-Wing UAV Model

Image 4.2 Dual Propulsion (on Wings) Fixed-Wing UAV Model

Image 4.3 Dual Propulsion – Front and Rear on a Designed UAV Model

Image 4.4 Rear Electric Propulsion System on a Fixed-Wing UAV

3. Usages

A Fixed-wing drone has many different applications. In terms of commercial applications, Fixed-wing drones are used for land mapping surveys, forest surveys and surveillance purposes. In military operations, a Medium Altitude Long Endurance (MALE) and High Altitude Long Endurance (HALE) Drones can be used for border surveillance. UAVs capable of carrying heavy payloads can be used for cross-border defence purposes. UAV aero models in the mini category can be used for certain defence related experiments and testing purposes. (The details of defence experiments and tests being carried out on Fixed-wing UAVs cannot be disclosed due to security reasons.)

4. Purpose/Payload-based Design

The purpose is basically an application for which the vehicle is going to be used for. With the change in purpose, the design, the flight time, the payload required, etc. are decided and set accordingly.

For a drone, there is a particular application planned to be executed. Based on the applications, different sensors are selected to be put on the drone. All these sensors, which are not a direct part of the drone, are considered payloads.

For example, a surveillance drone needs few external systems, such as a zoom camera, a gimbal for the camera operation and a couple of obstacle-avoidance sensors, to be used as payloads. Now all these have become payloads. Once the list of payloads is decided, it is good to go with the final product selection. The final product selection for the purchase and integration will help in getting details of the weight. With sensors, gimbal and camera along with all its cable, we get the weight, shape and size of the payload, movements required in the payload, and this helps in designing our drone in terms of dimensions and shape, selection of type of landing gears, control surfaces required in the aircraft (with/without flaps or any other), etc.

The selection of power systems also gets included under this purpose-based design i.e., electric or engine powered.

5. Control Surface Selection in the Design

Once your vehicle is designed as per the requirements, it's time for selecting and providing the control surfaces. In any aircraft, there are three main control surfaces: Aileron, Elevator and Rudder. The additional control surface includes flaps. With the shape of the wing and tail sections, few changes will occur in the control and operations of control surfaces.

V-tail (vee tail): In a V-tail design of an aircraft, the horizontal and vertical fins (like a conventional plane empennage) get converted into slanted tails of "V" shape. So, the rudder and elevator operation can be achieved with the mixing of this V-shaped control surface but will operate separately as elevator, when elevator stick moment is given, and as a rudder, when rudder stick moment is provided on the Remote Controller. This combined operation of the rudder and elevator with common control surfaces is known as Ruddervator.

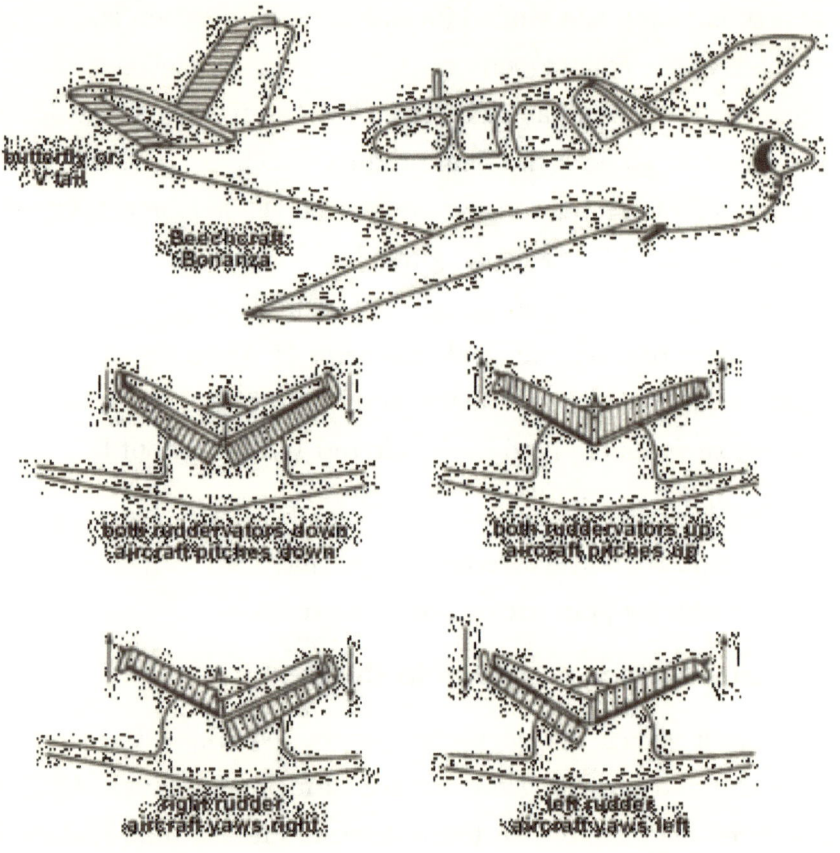

Image 4.5 V-tail Aircraft Tail Control Surface Operation[xxv]

The totally opposite configuration to this is A-tail (or an inverted V-tail), where the combination remains the same, apart from changes in the design of the aircraft.

Types of Unmanned Aerial Vehicles

Image 4.6 A-tail Aircraft

Similarly, for flying wing models, where the configuration of the elevator and aileron are mixed, they are known as elevon. These elevon-configurations have wings but no tail. Both pitch and roll moments are achieved with this configuration.

Image 4.7 A Flying Wing Drone[xxvi]

6. Purpose-based Selection of Materials, Equipment and Electronics

- All aircraft need different electronic and avionic systems.

- Electric-powered UAVs will require a brushless DC motor and an electronic speed controller for motor operation and control.
- Engine-powered UAVs are not operated using BLDC motors.

The following are the common electrical and electronic systems for engines as well as electric-powered Fixed-wing UAVs:

a) Drone control surface movements can be achieved using servo actuators.

b) The radio transmitter and receiver are used for the control of the drone from the ground.

c) An autopilot system is required for autonomous flight and control operations as per the requirements.

d) The ground control station is required for the control and monitoring of the autopilot operation. This can be in the form of a small display or a laptop.

e) Battery is used for powering the motor, servos, autopilot system and other electronics.

f) Telemetry system is basically a part of the autopilot system which is used for communication between the ground control system and the drone.

7. Engines/Motor

A propulsion system in a Fixed-wing drone can be an electric motor or a gasoline engine as per the design specifications.

Electric Propulsion

An electric motor is a BLDC motor connected with an electronic speed controller. This motor operates through a desired battery source and throttle can be controlled to the point of required speed of operation.

Image 4.8   A BLDC Motor for a Fixed-Wing UAV[xxvii]

Model Engines

Model engines are another type of power source that operates Fixed-wing drones. These engines are internal combustion engines and these can be a 2-stroke and 4-stroke engine. The most common engines used for RC model aircraft and Fixed-wing UAVs are glow engines/nitro engines and gas engines.

The nitro engines operate with methanol (mixed with oil) as a fuel. It differs greatly from the other internal combustion engines as it has got a glow plug instead of a spark plug. These glow plugs provide continuous heat via a fine platinum alloy filament that is a red hot glow which ignites the fuel and air mixtures inside the combustion chamber, whereas spark plugs produce rapid, repeated sparks.

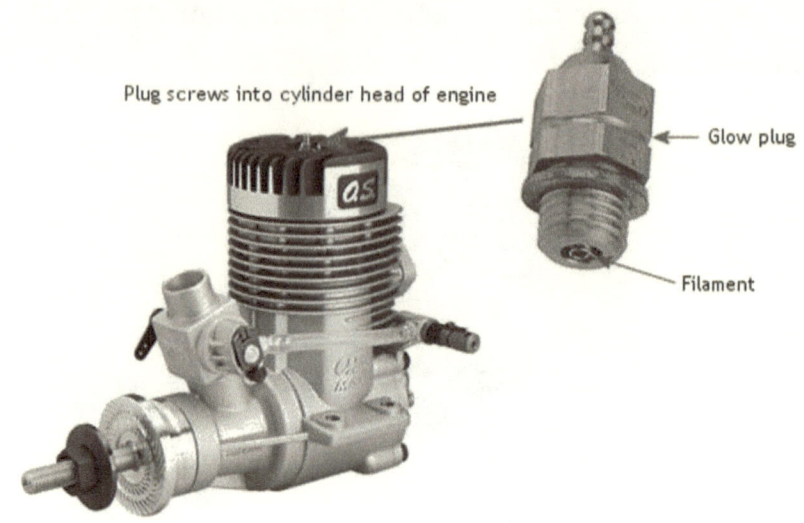

Image 4.9 Nitro Engine with Glow Plug[xxviii]

For this type of glow plug, an igniter is required at the time of cranking the engine as an initial supply of electricity to the glow plug. This is known as glow plug igniter. Glow plug igniters are battery operated and even rechargeable. It has to be placed on the glow plug and then the engine starts. Once the engine operation is normal, the igniter can be removed.

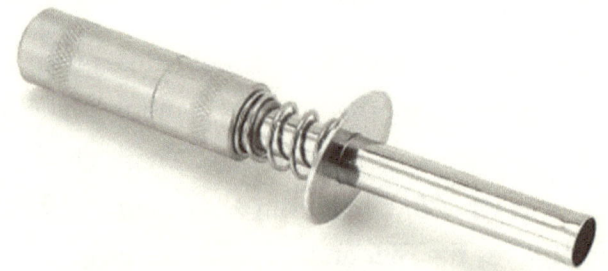

Image 4.10 Glow Plug Igniter

This type of engine uses methanol mixed with oil as its main fuel.

Another type of engine used is a gasoline engine that uses petrol/gasoline as a fuel for operation. The advantage of the petrol engine over a methanol engine is that petrol is cheaper in cost than methanol. The petrol engine burns cleaner whereas the glow engine leaves an oily residue. These engines do not require a glow plug igniter as it has an electronic igniter which automatically ignites the plug.

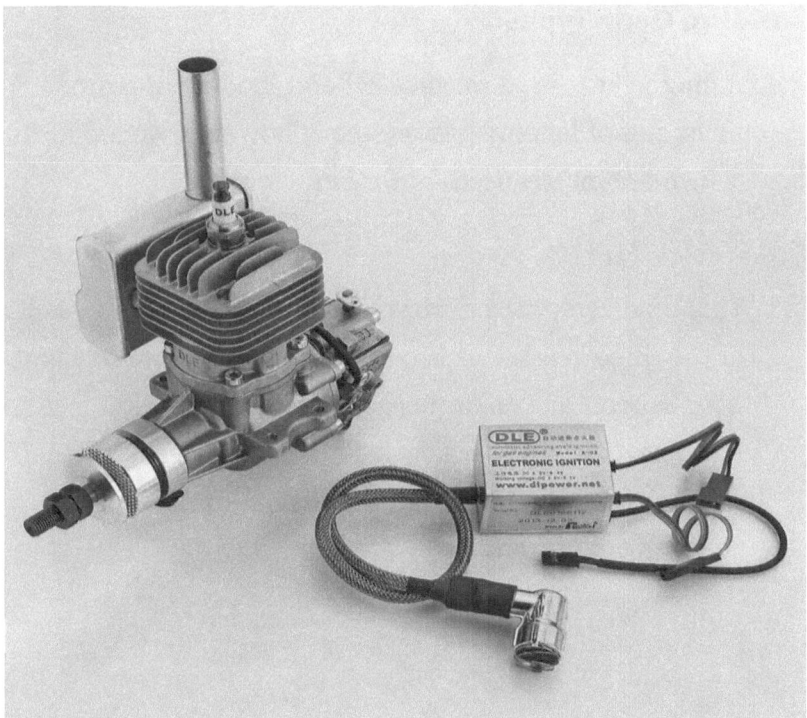

Image 4.11 An Image of a Gasoline Engine and an Electronic Igniter

8. Electronic Systems

The main electronic systems that will be required for the operation of the Fixed-wing drones are servo motor, power box, ESC, autopilot system, battery, ground control station, telemetry system for the drone, video telemetry system, etc. Additionally, all electronic systems are required for the desired payload operation.

9. Propellers

The description of the propeller is provided in chapter 5.

In Fixed-wing UAVs, apart from carbon fibre, plastic and glass fibre nylon propellers, wooden propellers are also used in engine models.

10. Landing Gear Types

The landing gear is used for take-off and landing of a drone. There are many types of landing gear systems; however, we are looking at those which are mostly used for drones.

a)   Tricycle type

This type comprises a tricycle layout for landing wheels. Out of the three gears, one is placed in the front near or under the nose section and the other two are at the back joined to the fuselage. The front wheel is known as the nose wheel. This nose wheel is also used for the change in the direction of the vehicle while taxiing.

Image 4.12   Tricycle Configuration[xxix]

b) Tail Dragger

In this type of configuration, the main landing gear is attached slightly before the CG point. Due to this, the other small gear is required to be attached to the tail portion.

Image 4.13   Cessna 150 Taildragger[xxx]

11. Take-off and Landing Methods

For any type of Unmanned Aerial Vehicle, launch and recovery is crucial and requires certain skills. A Fixed-wing drone requires a certain minimum controllable speed when it is in the air after take-off. To achieve that minimum speed, a craft needs some stretch over which it can achieve the speed and take-off easily.

For a Fixed-wing drone with traditional landing gears, a good runway for take-off and landing, preferably straight and plain strip, is required. Many a time, it's not always a strip and it's an open ground from where we take-off the drone. However, it is recommended to have a levelled path.

Image 4.14 A Fixed-Wing Drone that Requires Runway[xxxi]

Some drones do not have any landing gear for take-off and landing operations. Such drones need to be hand launched.

Image 4.15 Hand launching of Puma Drone (Credit: The USA Army)[xxxii]

Many Fixed-wing UAVs take-off with the help of rocket assisted take-off systems, catapult launches and bungee launches.

The catapult launching system provides external assistance to take-off Fixed-wing drones in locations where space is limited for take-off and may not even have a dedicated runway. Pneumatic and hydraulic launching systems require power supply for operation, whereas bungee-assist makes use of the energy that is stored in highly elastic bungee cords to launch UAVs.

While most of the drones land on their landing gears, few have a belly landing approach. In such cases, it is good to land on the grass so that it does not damage any part of your plane.

Image 4.16   Belly Landing on the Grass Ground

Some drones have parachutes that need to be triggered during landing. Large drones used in the Navy may have net recovery.

Image 4.17 A Pioneer I Remotely-Piloted Vehicle (RPV) is Caught in a Recovery Net Erected on the Stern of the Battleship USS Iowa (BB 61), December 1986.[xxxiii]

12. Assembly and Integration

    a) The main assembly of a Fixed-wing vehicle starts with the collection of already fabricated parts.

    b) Check whether all the parts are fine to be assembled. In case of any changes required, please do it.

    c) Landing gears should be attached first so that it becomes easy to assemble the other things.

    d) Attach all the actuators/servo. Attach servo to wings for the aileron and flaps, if any.

e) Attach the engine/motor to the front or rear mount.

f) For the engine, do the ignition connections, tighten servo linkages and fuel pipe connections with the tank and engine.

g) In the case of the electric motor, do all the ESC wirings.

h) Connect to the autopilot system and configure the system properly.

i) Connect the receiver and telemetry system with the autopilot system.

j) If the wing is detachable, attach it at the end.

k) If the wing is to be fixed, attach it once all the wiring set up is done and tested.

l) Check all control surface operations in manual and auto-mode.

m) Different types of wings may need different types of configuration setup parameters to be done in the autopilot system.

n) Attach the propeller, hatch, battery and all the remaining accessories, including cowl, and check the CG balancing.

o) If there is any CG unbalancing, analyse and perform certain product placement changes and getit balanced.

p) In the case of an electric motor, attach the propeller and operate it. Check whether the thrust direction is correct.

q) In the case of the engine model, attach the propeller and tighten it properly.

r) Fill up the fuel and go for engine operation testing.

s) Once everything seems ready, the drone is ready for manual testing and autopilot tuning.

4.2 VTOL Drones

1. Introduction

VTOL stands for Vertical Take-off and Landing. As the name suggests, it is a type of drone that takes-off vertically and lands vertically. It is necessary to not confuse this type of drone with Multirotor systems. It is quite correct that a Multirotor system also does take-off and lands vertically. However, here the drones are a combination of the features of a Fixed-wing drone as well as Multirotor systems. So, it will basically be a vertical take-off (with Multirotor mechanism) and horizontal forward transition (with Fixed-wing mechanism).

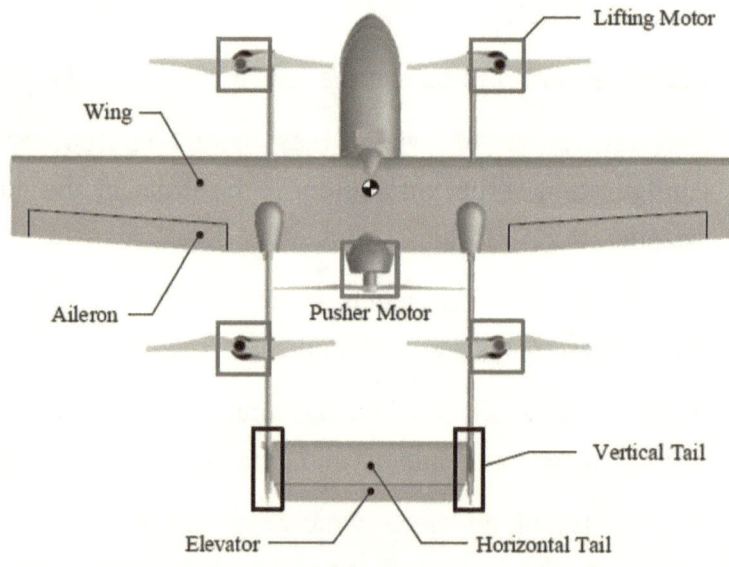

Image 4.18 A Design of a VTOL UAV[xxxiv]

As shown in the above image of a designed VTOL drone, there are four motors that are named as Lift Motors. When these four motors are operated, they work as a Quadcopter. This Quadcopter operation part is for a short period of time, mainly required for take-off and landing. Sometimes, this Quadcopter operation may be executed when there is a requirement of hovering during a mission.

The other part we can observe resembles a Fixed-wing plane model with aileron, elevator and rudder as the main control surfaces and a pusher motor for horizontal transition. So, this operates as a Fixed-wing plane.

Once the take-off is done and the drone is at its desired altitude, the drone switches from vertical motor operation to pusher motor operation for horizontal transition. The same happens when the drone returns from the mission for landing.

2. Scale of UAV Based on Operation

The dimensions of a drone are calculated and designed based on the type of application. This calculation includes weight estimation as well as other parameters of design that can affect or improve flight performance.

3. Usages

Like a Fixed-wing plane and a Multirotor system, this drone too can be used for several applications like mapping and surveillance. In mapping, it can be used for fast photogrammetry purposes as it will cover a large area in a short time and at any typical location with no need of a runway but with a proper flat surface. This drone can also be used for payload delivery like medical emergency delivery and first-aid. VTOLs also have the capabilities to work on border inspections, mining surveys, power line inspections, infrastructure inspections, railways and road inspections.

4. Purpose/Payload-based Design

The payload-based design basically implies that once the type of payload to be attached is finalised, for example, if the payload is a big cartoon box, drone can be designed accordingly and if the payload is a camera or a small sensor, then the design can be implemented as a normal model.

For the purpose of carrying a small camera or a sensor, a Fixed-wing drone can also be converted into VTOL by doing proper calculation and adding up the Multirotor system setup on the Fixed-wing. This can be done in many different ways.

5. Purpose-based Selection of Materials, Equipment and Electronics

For VTOL configuration, different materials may be required for different systems viz. Fixed-wing and Multirotor. This basically depends on the weight and payload capacity as this is directly proportional to the strength required.

Most of the VTOLs weighing more than 8 kg will probably be fully made of carbon or glass fibre and kevlar. In order to reduce the weight, the arms for multirotor can be made up of carbon fibre attached with the aircraft body made up of balsawood.

All aircraft need different electronic and avionic systems.

Hybrid VTOL can be an electric-powered UAV with an engine for horizontal transition, or it can be a fully electric-powered system. A hybrid system of electric and engine power will require a brushless DC motor and an electronic speed controller to lift motor operation and control. For the pusher thrust, an engine and its accessories will be required.

In the case of fully electric VTOL (even with vectoring motor system), BLDC motor paired up with ESCs will be required.

Common electronic systems for engines as well as electric-powered VTOL UAVs:

a) Drone control surface movements can be achieved using actuators.

b) The radio transmitter and receiver are used for the control of the drone from the ground.

c) An autopilot system is required for autonomous flight and control operations as per the requirements.

d) The ground control station is required for the control and monitoring of the autopilot operation. This can be in the form of a small display or a laptop.

e) Battery for powering the motor, servos, autopilot system and other electronics.

f) Telemetry system is basically a part of autopilot system, used for communication between ground control system and the drone.

6. Motor/Engine

The configuration of a fully electric system will require having all the BLDC motors for lift motors and pusher motors. Motor size and KV rating will vary along with the propeller and ESC capacity for pusher motors and lift motors as the thrust calculation for Fixed-wing configuration and Multirotor will vary here.

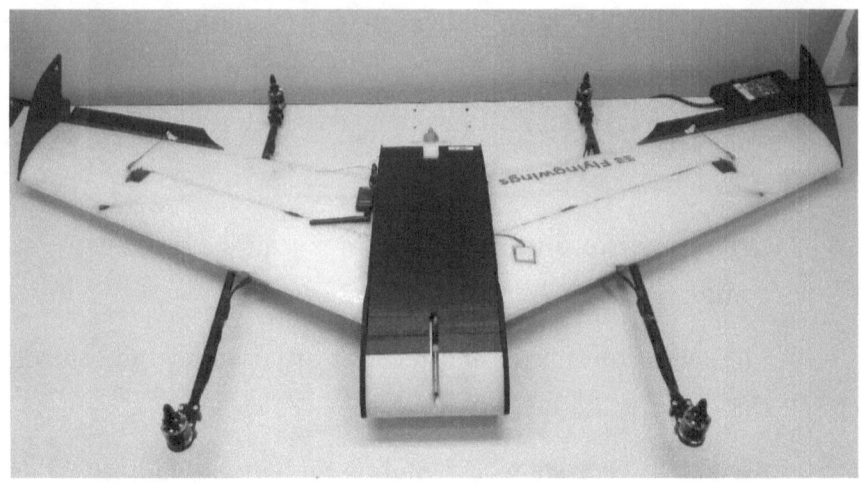

Image 4.19 Fixed-Wing Aircraft with Forward and Vertical BLDC Rotors[xxxv]

In the below shown configuration, instead of rotating the engines/motors, the whole wing rotates in order to change the direction of thrust and lift the drone vertically. Once it is in the air, the transition for the rotation of the wing is carried out so that the aircraft can move forward horizontally.

Image 4.20 Airplane with In-flight Rotation of Wings and Rotors[xxxvi]

Image 4.21 Airplane with Rotation of Engines During Flight[xxxvii] (Credit – Looper5920)

The above shown image is of a hybrid VTOL aircraft that has thrust vectoring system. The attached engines are the sole propulsion system that will generate thrust as well as lift. The thrust is generated by positioning the engines vertically, and the lift is generated by rotating the engines by 90° upwards (horizontal).

The above configurations will either need engines that are shown in the Fixed-wing vehicle description and/or BLDC motors to carry out the operations.

7. Electronic Systems

The main electronic systems that will be required for the operation of the VTOL drone are servo motor, power box, ESC, autopilot system, battery, ground control station, telemetry system for the drone, video telemetry system, etc. Additionally, all electronic systems are required for the operation of the payload.

8. Landing Gear Types

Landing gear is used for take-off and landing of a drone. In Fixed-wing drones, two types of landing gears are used, as mentioned above. However, for VTOL drones, it is not mandatory to have the wheels for landing gears. It can simply be landing legs like a typical Multirotor system.

Image 4.22 A Dual Leg Type Landing Gear of a VTOL Drone

The above image describes the feature of a landing gear system on a VTOL drone. This drone has no wheels to roll on the runway. The advantage of having wheels with landing gears is that while carrying out certain missions, there may be the availability of the runway and the drone can be taken off as a Fixed-wing drone. Another good reason to have wheels attached to the landing gear is that in case of accidental failure of any of the lift motors, the drone can be operated as a Fixed-wing drone if not as a VTOL UAV.

9. Assembly and Integration

a) The main assembly of a Fixed-wing vehicle starts with the collection of already fabricated parts.

b) Check whether all the parts are fine to be assembled. In case of any changes required, please do it.

c) Landing gears should be attached first so that it becomes easy to assemble the other things.

d) Attach all the actuators/servo. Attach servo to wings for the aileron and flaps, if any.

e) Attach the engine/motor to the front or rear mount for horizontal transition.

f) For the engine, do the ignition connections, tighten servo linkages and fuel pipe connections with the tank and engine.

g) In the case of an electric motor, do all the ESC wirings.

h) Assemble the electronics of two or three or four motors for vertical take-off and landing operation.

i) Connect to the autopilot system and configure the system properly.

j) Connect the receiver and telemetry system with the autopilot system.

k) If the wing is detachable, attach it at the end.

l) If the wing is to be fixed, attach it once all the wiring set up is done and tested.

m) Check all control surface operations in manual and auto mode.

n) Different types of wings may need different types of configuration setup parameters to be done in the autopilot system.

o) For VTOL, there may be different parameter settings required to be set up for the correct operation.

p) Attach propeller, hatch, battery and all the remaining accessories including cowl and check the CG balancing.

q) If there is any CG unbalancing, try to balance the vehicle by changing the battery position.

r) In the case of an electric motor, attach the propeller and operate it. Check whether the thrust direction is correct.

s) In the case of the engine model, attach the propeller and tighten it properly.

t) Fill up the fuel and go for engine operation testing.

u) Once everything seems ready, the drone is ready for manual testing and autopilot tuning.

10. Take-off Methods and Landing Methods

Take-off and landing of most of the VTOL drones are similar to a Multirotor system.

4.3 Helicopter Drones

A helicopter drone is a vehicle that operates like a traditional helicopter system; however, it is quite different from Fixed-wing drones as it has got the hovering capabilities. The technical parameters, aerodynamics and flight operations are totally different compared to RC aircraft. Helicopter drone is a type of drone whose operation comes in between the Multirotor system and the Fixed-wing system as well as it is a typical system than the other two. Apart from hovering advantage, helicopters can take-off and land vertically.

Helicopter drones have long endurance and higher payload capacity. They can be used in aerial LIDAR-based surveys. Single rotor UAVs are also used in military surveillance operations. For propulsion, it is good to use the nitro or gas engine. This will also increase the payload capacity.

Types of Unmanned Aerial Vehicles

Image 4.23 A Camcopter Helicopter Drone[xxxviii]

The main electronic systems that will be required for the operation of a helicopter drone are servo motor, power box, ESC, autopilot system, battery, ground control station, telemetry system for the drone, video telemetry system, etc.

Note that the Multirotor system too comes under the types of UAVs, as shown under the classification chart at the beginning of the chapter. However, as the focus of this book is more on trending drone i.e. Multirotor systems, it is not covered under this chapter; instead, a separate chapter (5) has been designed to cover information on Multirotor UAVs.

Chapter Insight

As Unmanned Vehicles are not limited to air, Unmanned Aerial Vehicles too are not limited to Quadcopter or Multirotor systems. It can be a Fixed-wing or even hybrid VTOL that flies with some sort of aero-dynamic effect through its wings or propellers. The selection of the vehicle should be done with the consideration of the application requirement. For example, in case of the requirement of hovering, a Fixed-wing drone cannot be used. So, choose wisely.

CHAPTER 5
Multirotor UAV Systems

5.1 Introduction

Multirotor UAV is a vehicle that has multiple motors or rotors, in horizontal positions, for control of its operation and motions. The number of motors varies from two to eight or even more, depending on the requirements and set configuration. So, with the change in the number of motors, the name of the drone also changes. The one with four motor configurations is known as Quadcopter/Quadrotor. Another one with six motors is known as Hexacopter. The reason behind the word "Multirotor" is to express and address all the multiple rotor mounted platforms and vehicles. As they all have the same physics of flying and operation.

The horizontally placed rotors produce vertical thrust. The concept of thrust production is the same as a helicopter. However, there is no tail shape or portion available like it is there in a helicopter. Vehicle motion is controlled by varying the relative speed of each rotor. Additionally, when compared to a helicopter, here the pitch of all the propellers are the same and fixed.

Basics of Unmanned Aerial Vehicles

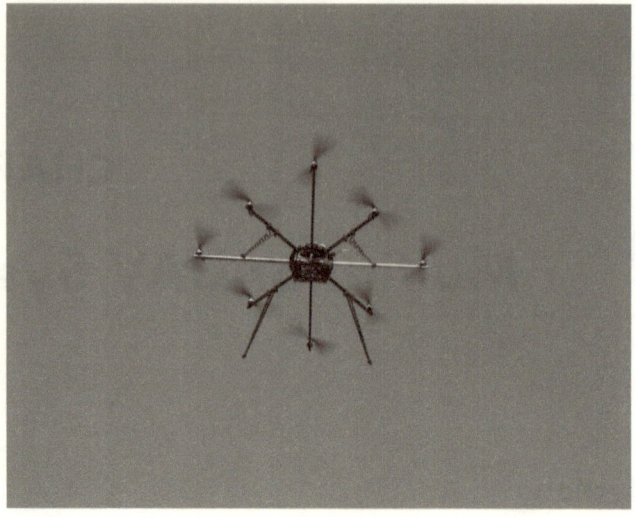

Image 5.1 A Multirotor System[xxxix] (Credit: Hans Braxmeier)

5.2 Multirotor Configurations

Multirotors are categorised on the basis of the number of motors. The four motors will be considered as Quadcopter. The types of configurations that are popular have been described below:

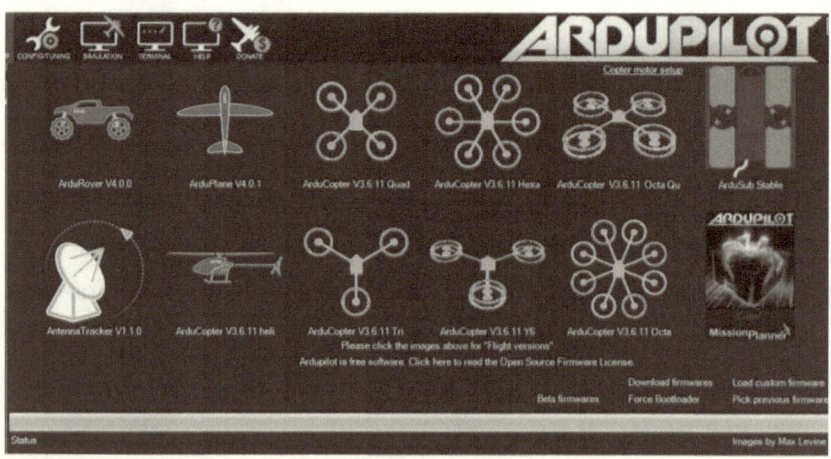

Image 5.2 Screenshot of Mission Planner Compatible Vehicle Selection Configuration

a) **Bicopter**

Bicopters have two motors but are quite unstable in the beginning. This type of drone requires much more stabilisation which can be achieved by some performing some precise parameter settings. This drone has not so defined applications so far, under the micro and nano category.

b) **Tricopter**

Tricopters are mostly in Y shape. As per the 360 degree rule, the arms are designed at 120 degree apart from each other. In tricopter, servo is connected to the tail motor. This servo provides angular movement to the motor with our controls, in order to achieve the desired yaw moment.

Image 5.3 A Tricopter Drone model[xl]

c) **Y6**

When a tricopter configuration includes a coaxial motor arrangement on the drone, it turns out to be a Y6 tricopter. The difference in the configuration is at the point of yaw applied to the drone. In tricopter, the drone yaws with the angular movement of the tail rotor, whereas in Y6, the coaxial motor arrangement eliminates the gyroscopic effect, allowing the drone to yaw easily and normally. A hexacopter with the same configuration will weigh more than a Y6 copter.

Image 5.4 A Y6 Drone Model[xli]

d) **Quadcopter**

Quadcopters have four rotors or motors and can be in different shapes. The most common shape is the "X" type. The other shapes are "+" and "H" configurations. In Quadcopter, the rotors are 90 degrees apart from each other.

Below shown are the images of the X, + and H configurations. The difference between these three configurations is just the shape as per the requirement. Certain custom made designs may allow for more space to put in all the electronics and payloads.

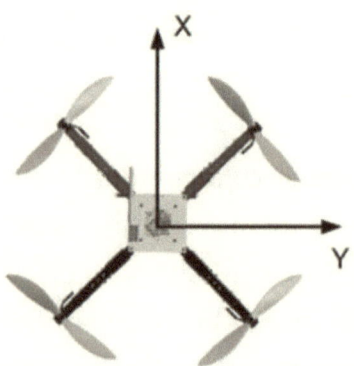

Image 5.5 Quadcopter "X" Configuration Example[xlii]

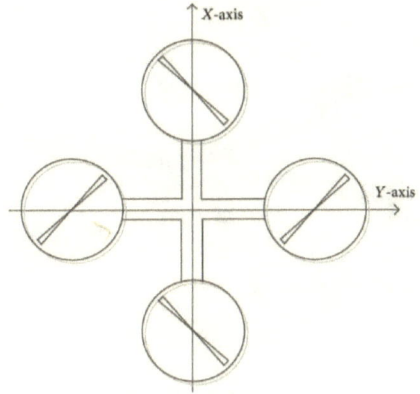

Image 5.6   Quadcopter '+' Configuration[xliii]

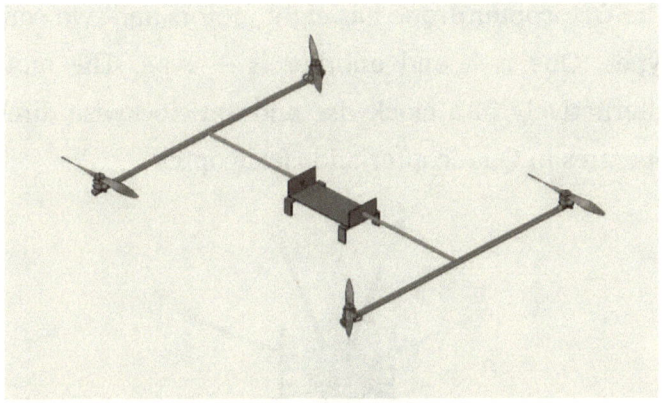

Image 5.7   H-Type Drone (3D Design) Isometric View

e) **Hexacopter**

The Hexacopter drone has six motors and two configuration types. One is X and another is + type. X-Type is the most common in hexacopter drones. The motors are placed at a 60-degree angle from each other and rotate alternatively in a clockwise and anticlockwise direction (as shown in the configuration image).

Basics of Unmanned Aerial Vehicles

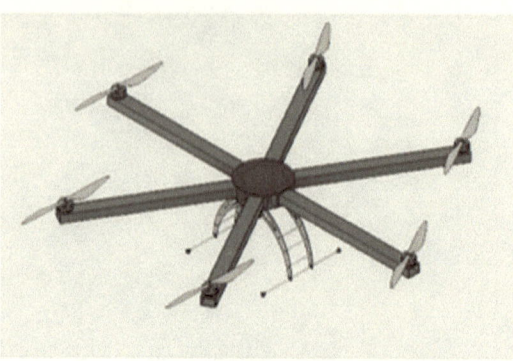

Image 5.8   X type Hexacopter 3D Design and Configuration

f) Octocopter

The Octocopter drone has eight motors and two configuration types. One is X and another is + type. The motors rotate alternatively in a clockwise and anticlockwise direction as it operates in Quadcopter and Hexacopter.

Image 5.9   Octacopter[xliv]

g) OctoQuad

X8 is also known as octo-quad. This is a Quadcopter configuration with the coaxial motor arrangement. The advantage of this model is that the weight of large octocopter frame decreases as there are only four arms, the size also decreases but the weight of additional motors and power consumption increases. The endurance provided by this drone can be high at a smaller size than a hexacopter or an octocopter for a similar payload carrying capacity.

Image 5.10   OctaQuad[xlv]

The advantage of hexacopter and octocopter configuration over Quadcopter is that they carry more weight as the payload capacity increases. However, this is also a matter of fact that the craft itself becomes heavy due to the added number of motors and a bigger battery. The larger the size, the more will be the number of motors, ESCs, propellers and larger capacity battery. It will add up everything to the total cost; so, the bigger configurations are expensive too.

An additional point which is important to note that when compared with Quadcopter, Hexacopter and Octocopters are considered to be safer at the time of any sort of failure, viz. one motor or ESC stops operating or a propeller breaks/comes out, as it will provide some short period of time to land it safely instead of getting crashed like a Quadcopter.

5.3 Usages

The drone can be used for any of the applications explained in chapter 12. An application can also be based on a new idea, and this will change the usage of drones, the payload required and many other related parameters. The decision about the application for which the drone will be used matters a lot as the total design of a drone is dependent on the application.

5.4 Purpose/Payload-based Design

As the payload has now been decided, it is easier to identify which product to go for as a sensor or any product related to your application. It is necessary to check the technical specifications, dimensions, weight and other power connection parameters for the sensory payload. Depending on the size and weight, a Multirotor configuration can be selected out of Tricopter, Quadcopter, Hexacopter, Octocopter, Octoquad, etc. as per the requirement. Also, the choice can be made in terms of "X-Type" or "+ Type" configuration for Quadcopter and Octoquad, as per the configurations explained above and based on the requirements.

For example, a photography drone needs a few obstacle avoidance sensors, a camera and a gimbal for the camera operation. Now, all these becomes a payload. Once the list of payloads is decided, it is good to go with the final product selection. The final product selection for the purchase and integration will help in getting details of the weight. With sensors, gimbal and camera along with all its cable, we get the weight, shape and size of the payload, movements required in the payload, etc. and these details help in designing our drone in terms of configuration selection, its dimensions and shape.

5.5 Scale of UAV-based on Operation

The design of a Multirotor system is totally dependent on parameters like weight expectation, payload carrying capacity, endurance requirement, range requirement and the cost, directly or indirectly. As per the requirements in the aforesaid parameters, the scale of a UAV can be decided.

> As per the DGCA norms in India, the following are the five categories of drones:
>
> a)   Nano UAV: It weighs less than or equal to 250 grams.
>
> b)   Micro UAV: It weighs between 250 grams and 2 kg.
>
> c)   Small UAV: It weighs between 2 kg and 25 kg.
>
> d)   Medium UAV: It weighs between 25 kg and 150 kg.
>
> e)   Large UAV: It weighs above 150 kg.
>
> As per Indian standards, these categories are meaningful and should be taken into consideration before designing a vehicle so that it becomes easier to follow DGCA norms and procedures for further permissions and modifications, if applicable.

Nano-sized drones can mostly be used in defence for secret surveillance purposes with drones like nano Hummingbird and black hornets. Apart from such applications, a nano-sized Quadcopter will not be of much use in any applications.

Drones with a weight under 250 grams can maximum allow the coreless motor to generate thrust instead of BLDC motors. A typical coreless motor weighs around 2 to 5 grams and can rotate at around 50,000 RPM. This configuration can mainly be used in nano Quadcopter and not for higher-end versions like hexacopter or octocopter in nano size. The scale of such drones will not be more than 30cm (motor to motor) and can carry a single cell battery of 200 or 300 mAh maximum.

Micro drone can be designed when there are lightweight payload-carrying applications such as photography, inspections, surveillance, lightweight payload drop, etc. The reason behind this is the payload weight is around 250 grams for the sensors that are used in agriculture surveys, solar (thermal) inspections, land mapping, highway and corridor surveys, etc.

5.6 Purpose-based Selection of Materials and Electronics

5.6.1 Materials

The airframe of a drone is the main physical structure which should have enough strength and be able to load all the electronics, payloads and other components, as per the requirements. Single or multiple materials may be required in order to build a frame and assemble it, turning it into a high strength rigid structure that can withstand vibrations and do not transmit it further into the frames so that it does not impact the performance of the vehicle and fly stably. Different materials that can be used for the development of drones are plastic, aluminium, carbon and glass fibre composites.

Plastic is used for certain low-end drones and rarely used in the development of drones that carry a heavy payload as it cannot withstand more weight. It has good durability and high strength to weight ratio and lower conductivity. It is used for drones that are under 2 kg weight categories.

Whenever a drone is to be developed for application purposes, it is good to avoid frame selection or development with plastic material.

Carbon composites are also referred to as Carbon Fibre Reinforced Polymer.

The combination of carbon fibres and thermosetting resins contains high strength that enhances the durability of the drone. This composite also reduces the overall weight of the vehicle compared to other materials like aluminium. Carbon composite frames are useful for any category of drones as they can withstand heavy payload capacity.

Image 5.11 Carbon fiber frame for a Quadcopter

Glass fibre composites to be high strength, light-weight and robust. It also has good dimensional stability, high resistance to temperature and high electrical insulation.

Image 5.12 Glass Fibre Frame for a Quadcopter[xlvi]

Wooden and aluminium frames can also be used for drone framing; however, wood is always going to have limited strength capacity and if not, it is going to be a heavy frame. Aluminium is used in many drones.

Image 5.13 Wooden Frame for a Quadcopter[xlvii]

Overall, composite frames are preferable for the development of drones as they have less frame weight and are capable of carrying more loads. However, composite materials are expensive, have higher maintenance cost but need low maintenance. Another advantage of composite material is that it can be in rectangle pipe, square pipe, rounded pipe and flat plate form that can best fit in the arms and landing gears of the drone.

5.6.2 Motor

The BLDC motor selection is done based on the calculations of thrust requirements and related parameters.

5.6.3 Electrical/Electronic Systems

Multirotor drones will include motors, electronic speed controller, autopilot system/flight controller board, telemetry system, remote controller and receiver, drone and remote controller batteries, etc. The details of all the electrical/electronics are provided in chapter 6.

5.6.4 Propellers

Propellers are basically blades that are shaped in wing form which converts rotary motion from an engine to airflow that will push the drone upward producing lift, like a wing. The pitch angle of the propeller defines the possible displacement of the vehicle. For helicopters and certain aircraft, this pitch is variable and needs to be changed for certain control operations.

Like aerofoils, these propellers also have a leading edge and trailing and based on that it gets attached to the motor for its desired clockwise and anticlockwise rotations.

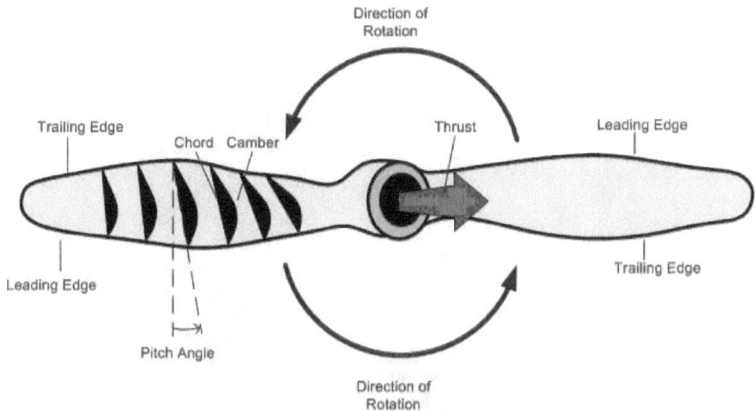

Image 5.14 Propeller Pitch and Other Parameters Defining[xlviii]

For all commercial drones, the propeller pitch is fixed at whatever value. Once fixed, this is not variable while the drone is in operation. Short-length propellers generate less lift and require less energy to reach any desired speed. They do not make the drone fly stable and hover at one position. Long-length propellers generate more lift and are quite stable under hovering conditions. With the decrease in the Kv rating of the motor, the length of the propeller increases.

A drone propeller may have two, three or four blades based on the design requirement. With more number of blades, the efficiency increases and also the drag. For any number of motors on the same drone, it is necessary to have the same blade propeller of the same size on that particular drone. For example, a Quadcopter, with one 1045 2-bladed propeller made of plastic, should have the other three of the same size 1045, 2-bladed, same quality and made up of plastic. Propellers are specified with a measurement X*Y. X means for propeller length and Y means for propeller pitch. For example, the Gemfan 9045 is a propeller with 9-inches length and 4.5-inches of pitches.

The material of a propeller is mostly carbon fibre and plastic for Multirotors.

Image 5.15   2-bladed Plastic Propeller

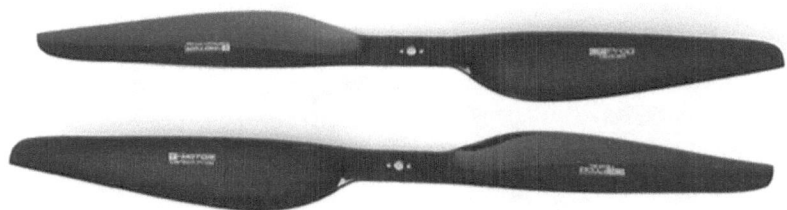

Image 5.16 2-bladed Carbon Fibre Propeller

The propellers are sharp and can cause injury. It is necessary to take utmost care while working with them. Follow the instructions set in the annexure to learn about the care to be taken while working with the propellers.

5.6.5 Landing Gear Types

Landing gears are the legs on which the drone rests while it is on the ground, at the time of take-off and landing. Based on the whole drone body, the material of the landing gear is selected. If the whole drone is made up of carbon fibre, the chances are higher that the landing gears would also be made up of the same material. Landing gears are selected on the basis of the required ground clearance. This ground clearance calculation depends on the payload defined to be carried. Some gimbals are longer in size, and some drones have a fixed mount camera. So, the one that has a gimbal needs to have more ground clearance than the one with a fixed mounted camera on the drone.

Types of Landing Gear

a) Fixed Landing Gears

The fixed-landing gears are attached to the drone's body and remain fixed in their position.

Image 5.17 Landing Gear for a Customised Drone

b) Retractable Landing Gears

Retractable landing gears have servo actuators that control and operate the landing gears. With the help of these servos, attached on both sides of the landing gears, the landing gears get retracted parallel to the width of the drone.

Image 5.18 Retractable Landing Gears and the Mechanism Used

5.6.6 Take-off Methods and Landing Methods

Multirotor drones' take-off and landing are quite simple compared to the take-off and landing of any Fixed-wing drone. The main advantage of a Multirotor system is that it does not need any runway for take-off or landing the drone as these drones can take-off vertically from the resting position. Both take-off and landing can be done manually and automatically. If the drone is under testing, it is good to go for manual take-off and landing. Manual take-off and landing should be done slowly and gently. In case of retractable landing gears, it is important to retract the gears at the right time in order to avoid incidents.

5.7 Operation of Multirotor UAVs

The attitude of a Multirotor system is totally based on the operation of the rotors. The orientation and the motion of a drone depend on the thrust of the motors relative to the centre of gravity of the vehicle. Taking an example of a Quadcopter system, there are four motors viz. motor 1, 2, 3 and 4, as shown in the image below.

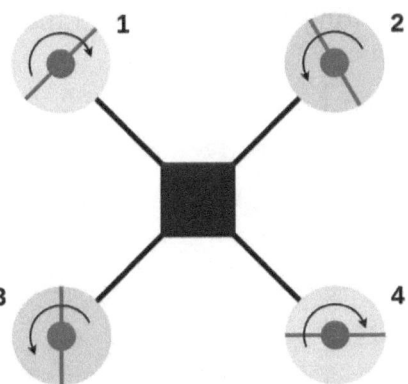

Image 5.19 X-type Quadcopter Configuration[xlix]

As shown in the image, motor 1 and 4 operate in a clockwise direction, whereas the other two operate in an anti-clockwise direction.

These four motors control the roll, pitch and yaw moments with the varying thrust produced by a set of the motors.

Check the image below for drone operation consideration.

Lifting: When all four motors are operated at the same speed or if rotor speed is increased gradually and equally, the drone takes off.

Pitching: In the pitch moment of the drone, the vehicle moves forward and backwards. So, when motors 3 and 4 run at a higher speed than motors 1 and 2, then the drone moves forward, whereas the backward movement is achieved with the motors 1 and 2 rotating at a higher speed than motors 3 and 4.

Rolling: The increase in RPM of the rotors 2 and 4 lets the drone move towards its right side and towards its left side if RPM of rotors 1 and 3 is higher than that of motors 2 and 4.

Yawing: In the yaw moment of the drone, the vehicle changes its heading/forward direction. When motors 2 and 3 run at a higher speed than motors 1 and 4, the drone yaws towards the left of the current heading direction, whereas the backward movement is achieved with the motors 1 and 4 rotating at a higher speed than motors 2 and 3.

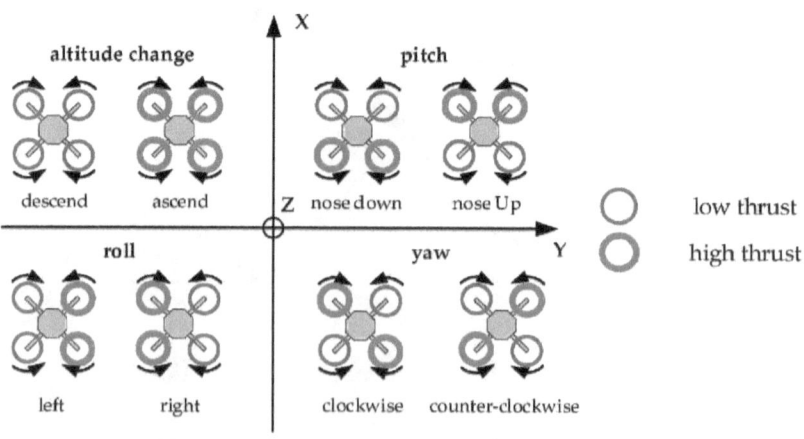

Image 5.20 Drone Orientation[1]

This 1, 2, 3 and 4 number of the motor depends on the autopilot configuration settings and varies with different types and brands of flight controller boards.

5.8 Assembly and Integration Steps

Once all materials are procured and all the parts are collected on the table, it is the time for the assembly of these parts. There should be a standard procedure to be followed for the assembly of your drone. However, this also varies with the various drones as not all drones provide enough assembly space that one may need to attach some electronics or payload.

In my personal point of view, the following steps are recommended for developing the drone from scratch. The provided steps can vary based on the different frame shapes, sizes and electronic placement constraints.

- Check the connectors required for the electronic speed controller, viz., bullet connectors and crimp connectors, are all there with all of your ESCs. If not, solder them. The cable of the power supply is to be soldered or connected with the power distribution board.

- Check whether the remote controller is linked to the receiver and whether the first-time setup is done.

- Check for all the required connectors that can operate your payload, including triggering them.

- Assemble the motor mounts on all the arms first; connect ESC cables to the motor. The wire should pass properly or it should be taped fine.

- Attach the power distribution plate on the centre plate of your drone

- The drone can be assembled starting with the centre base plate frame.

- First attach both the landing gears with the bottom centre plate.

- Then mount all the electronics that you want to put up between the top and bottom plates. This can include autopilot system, gimbal controller, ESC, receiver, telemetry, etc. Make all the required wiring connections of the autopilot system.

- Once done, you can connect to the battery and check if the connections are fine.

- If the operations are fine and as per the configuration settings required by the user and as per the autopilot parameters, you are good to go with the final assembly.

- There are chances that in the above testing, any of the motors may not operate or may operate in the opposite direction of our requirements. In that case, you will need to change the wiring between ESC and the motor of the same set and then test again.

- Do the same in case any other motor requires such changes in motor operation settings.

- Attach the top plate to the drone.

- Once all is fine, it is time to attach the propellers with the motors.

- Do the final configurations. Check on your flying application/ground control station.
- If everything is fine, it is good to go for the testing.

Chapter Insight

The selection of materials for drone frames and body should be strong enough to carry the payload, propulsions and other electronics. The recommended propeller size should fit the dimensions.

CHAPTER 6
On-Board Electrical Components

The selection and operation of electrical components that will be integrated on a drone are quite crucial and important as the wrong selection and improper operation of any of the electrical components may lead to a failure of the UAV. So, it is important to check the specifications and select products as per the requirements. Proper assembly and placement are also necessary parts in order to avoid any kind of failure.

6.1 Actuators

Servo Motor is the most common form of actuators used in all types of Fixed-wing vehicles as all control surface movements rely on it. An electric motor that is a DC motor in nature is in connection with a potentiometre, a controlling circuit and gear assembly mechanism, in order to achieve controlled and precise rotation of an object.

Working Principles

As mentioned above, a servo comprises of a gearbox, DC motor, potentiometre and a control circuit. The image below defines the assembly of a servo.

The position of gears is such that the first or the second gears are attached to the motor shaft, whereas the fourth or the final gears are attached with the potentiometre.

A series of pulses are sent through the signal line in order to control the servo motor. Once the motor rotates, the potentiometre also rotates and the voltage of the potentiometre is compared with the voltage applied/coming from the signal line. The difference between the two voltage signals should be zero. The controllers enable/activate the bridge circuit for counter-rotation for nullifying the voltage difference.

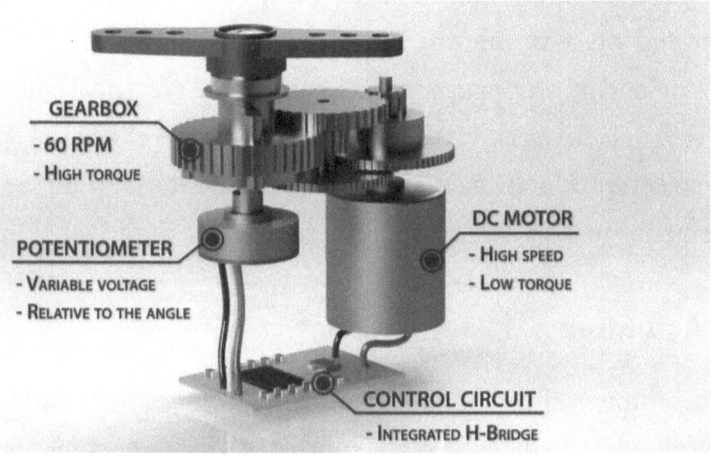

Image 6.1 Sub parts of a Servo[ii]

Coming to the part of controlling a servo motor, there are three wires that are outside for connection and operation. These three wires are positive, negative and signal. All three wires are in connection with the control circuit and are then internally connected to the motor and potentiometre.

PWM (Pulse Width Modulation) is the main principle on which servo motor operates. Here, the principle applies for controlling the motor operation duration by applying pulses. When the pulses are triggered, the speed force of the motor is converted into torque by gears. We know that WORK = FORCE × DISTANCE. In the DC motor, the force is less and the distance (speed) is high. In servo, the force is high and the distance is less. The potentiometre is connected to the output shaft of the servo, to calculate the angle and stop the DC motor at the required angle.[lii]

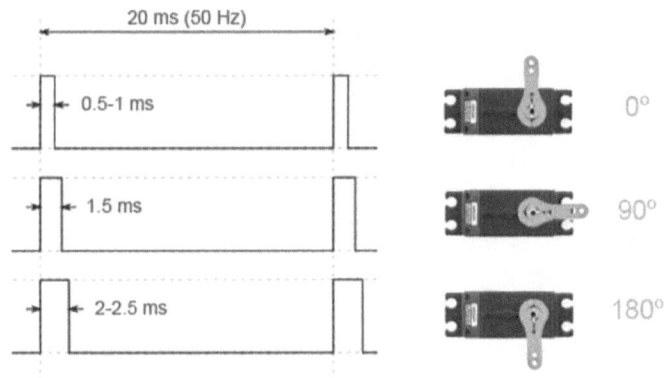

Image 6.2   Servo Operation[liii]

A servomotor is used to provide a motion or rotation deflection to a control surface. For example, as far as the Fixed-wing vehicle is considered, the rudder can be operated towards the left and right of the aircraft with the help of a servo. Similarly, it is used for the aileron, elevator and flap control surfaces for up and down motion.

6.2 BLDC Motor

The most important component for electric-powered Fixed-wing UAV Multirotor drones is a brushless DC motor. These motors make the drones fly. Nano drones may even work with a brushed DC motor. But we are more concerned about normal-sized drones used for certain major parts of applications and all these drones work on a brushless DC motor.

Brushless motor converts electrical energy into mechanical energy like any other motor. However, with BLDC, the resultant efficiency is higher, high torque to weight ratio, high reliability, reduced noise, gives good control and saves quite a much power when compared to the other counterparts. Apart from the above-mentioned advantages, the benefit of the brushless motor also includes less mechanical wear. So, it has a longer life span than any brushed DC motor.

Advantages

- Higher efficiency
- Less mechanical wear and tear
- Operation speed of BLDC motor is dependent on the frequency at which the current is supplied and not the voltage
- Can be operated at high speed
- Very less noise
- Longer lifespan

Disadvantages

- High cost
- Winding may get damaged if operated at a power higher than the prescribed limit

Image 6.3 A Brushless DC Motor

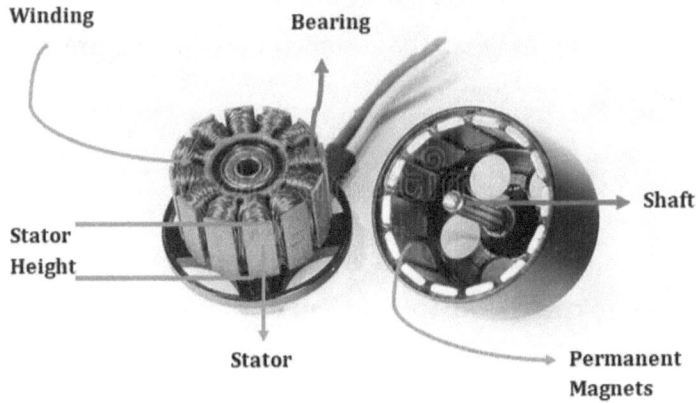

Image 6.4 Inside BLDC Motor

A BLDC motor in any drone is controlled by an electronic speed controller. There are three cables coming out of the motor, as shown in the above image. These three cables are phased-out and are connected to the electronic speed controller. The details of the electronic speed controller are given in the next topic. This mainly controls the rpm (revolution per minute) of the motor. Signals for motor control are generally sent by a pilot from the ground using remote controller throttle movements. In the case of the autopilot system, this rpm is controlled automatically as per the mission planned/requirements.

BLDC motors are available with Kv information. This information allows us to identify the RPM of the motor per 1 volt applied, without any load. For example, a 1,000 Kv motor with a 3 cell battery 11.1 V will provide a spin of approximately 10,500 RPM. The calculated requirement of thrust to weight ratio for your drone is the most important parameter for the selection of a motor. The current carrying capacity and limit is also an important factor. This current drawn by the motor also has a relation with ESC as it is controlling/operating the motor.

The brushless DC motors can run in a clockwise and anticlockwise direction. This happens when the connections of power are interchanged to get in the opposite direction.

6.3 Electronic Speed Controller

The electronic speed controller is a component connected to the BLDC motor in order to control the RPM of a motor remotely using a controller or by the command received from the flight controller.

ESCs are designed with different current capacities and this current value is dependent on the current drawn by the motor when it is operated at its full capacity. For example, if a motor has a peak current drawn of 35A, then an ESC with higher current value than the motor is selected.

Below is the image of an ESC of 30 Amp.

Image 6.5   A 30 Amp Electronic Speed Controller

As per the above image, we can see three sets of wires.

In the case of RC Aircraft: On the left side, the two symbol shows + (plus) and − (minus) sign. These two cables are power cables that connect with the battery to power up the ESC. The cables on the right-hand side with the A, B and C symbol get connected with the three-phase cables of the motor. This is for controlling the motor. Here gets the main aim fulfilled of motor control. But this is not possible without the use of the third set of cables. The third set of cables is a servo cable with red, white and black cables. This set gets connected to the receiver servo connector points. This is the main cable that is used to send PWM signals to the ESC, by operating the radio controller sticks as per the requirement.

In the case of Fixed-Wing UAV with Pixhawk Autopilot System: On the left side, the two symbol shows + (plus) and – (minus) sign. These two cables are power cables that connect with the power module of a flight control system (this mainly depends on the flight controller used). The cables on the right-hand side with the A, B and C symbol get connected with the three-phase cables of the motor. This is for controlling the motor. Here gets the main aim fulfilled of motor control. But this is not possible without the use of the third set of cables. The third set of cables is a servo cable with red, white and black cables. This set gets connected to the servo rail on the Pixhawk. For those who follow the AETR method, this pin gets connected to the third number pin on main outs. This is the main cable that is used to send PWM signals to the ESC by operating the radio controller sticks as per the requirement or by the command received by the flight controller as per the mission planned.

Image 6.6   Pixhawk Servo Rail

In the case of a Multirotor system with Pixhawk autopilot system: On the left side, the two symbol shows + (plus) and – (minus) sign. These two cables are power cables that connect with the power distribution board. PDB in Multirotor system is used as there are multiple output operations with a single power source.

6.4 Power Distribution Board (PDB)

The typical power distribution board is used for connecting a single battery to the multiple ESC in a firm and better way. PDB consists of positive and negative terminal pads where the cables can be soldered as per the need. These terminals are internally connected. So, once soldered all the positive cables from the ESC on PDB positive terminal and all the negative cables from the ESC on PDB negative terminal, they will operate all together by connecting the battery to the PDB.

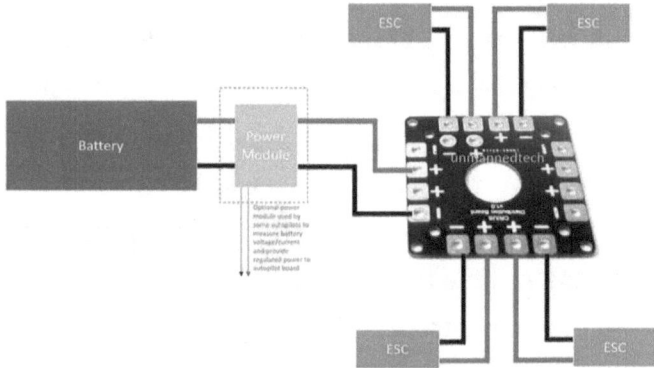

Image 6.7   A Diagram Showing Connection From ESC to Power Distribution Board and Battery[liv]

The cables on the right-hand side with A, B and C name marking (on ESC) connect with the three-phase cables of the motor. This is for controlling the motor. Here gets the main aim fulfilled of motor control. But this is not possible without the use of the third set of cables. The third set of cables is a servo cable with red, white and black cables. This set gets connected to the servo rail on the Pixhawk. For those who follow the AETR method, the right front motor gets connected on servo rail main out number 1. Accordingly, the other three get connected to the other three pins as shown in the image below. These are all the main cables that are used to send PWM signals to the ESC by operating the radio controller sticks as per the requirement or by the command received by the flight controller as per the mission planned.

Image 6.8 Servo Rail for ESC Wiring Connection on Pixhawk

30A BLDC ESC requires a standard 50-60Hz PWM signal from any remote control as a throttle input. Throttle speed is proportional to the width of the pulse. Maximum throttle position is user-programmable. In general, the throttle is set at zero for 1ms pulse width and full at 2ms pulse width.[lv]

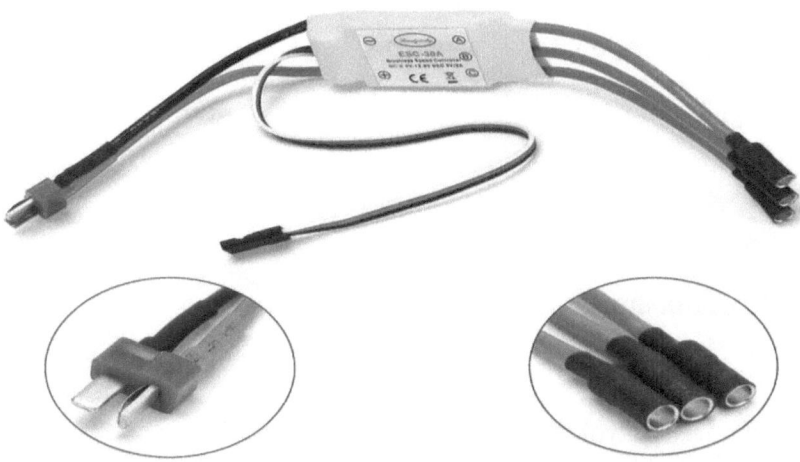

With connectors, avoid soldering problems.

Image 6.9 An Electronic Speed Controller Cables Explanation[lvi]

> ESC Cable Summary
>
> a) Cable connected to the deans-connector is for the power source.
>
> b) Crimp connection cable connects with the autopilot/receiver for controlling operation.
>
> c) Bullet connector cable connects with BLDC motor.

6.5 Battery

The most important subpart of the drone that supplies power throughout the system for the operation of the drone is a battery. Whether it is an engine-powered Fixed-wing drone or a Multirotor system, for all, some type of battery plays the main role for the drone operation. The battery in any drone can be used for the following purposes:

a) BLDC motor operations

b) Servo operations

c) Engine ignition

d) Gimbal operations

e) Camera triggering

f) Sensor operations

g) Flight controller operations

h) Remote controller operations

i) Operate miscellaneous other electronics

The following are the different types of batteries used in drone industries:

a) Lithium Polymer Battery (LiPo)

The most often used batteries in today's drones and different types of UAVs are Lithium Polymer Batteries, made up of the polymer electrolyte and are rechargeable.

LiPo batteries are being used by many industries, mainly due to their advantage in dimensions and weight. These are light in weight and can be designed and developed in any shape or size as per the requirements. The most common usage of LiPo batteries is in our mobile phones, tablets, laptops, power banks, a few peripheral devices, etc. Over the past few years, its usage has also increased in the field of electric vehicles, drones and UAVs. In UAVs and drones, the weight impact is the highest. The lesser the weight, the higher the endurance one can achieve with the required power setup integrated on a drone.[lvii]

Advantages

1. Light-weight
2. No shape or size constraint
3. High power holding capacity
4. Higher charge and discharge rate

A single cell of any LiPo battery contains a nominal voltage of 3.7V. When this cell is fully charged, the voltage will be 4.2V.

Image 6.10   Single LiPo Cell

The LiPo battery will have a combination of different cells in its pack. So, a three-cell LiPo battery will have three cells inside it. This three-cell pack will have a nominal voltage of 3.7*3 = 11.1 V. Under 100% charged condition, this will be a 12.6 V battery (4.2 V/cell).

Coming to the usage of a LiPo battery, the burst discharge and continuous discharge rating are always provided in a branded product. This will allow one to identify the maximum consumption. This calculation also needs to take battery capacity into consideration. Look at all the parameters provided on the battery pack/box or in a specification sheet.

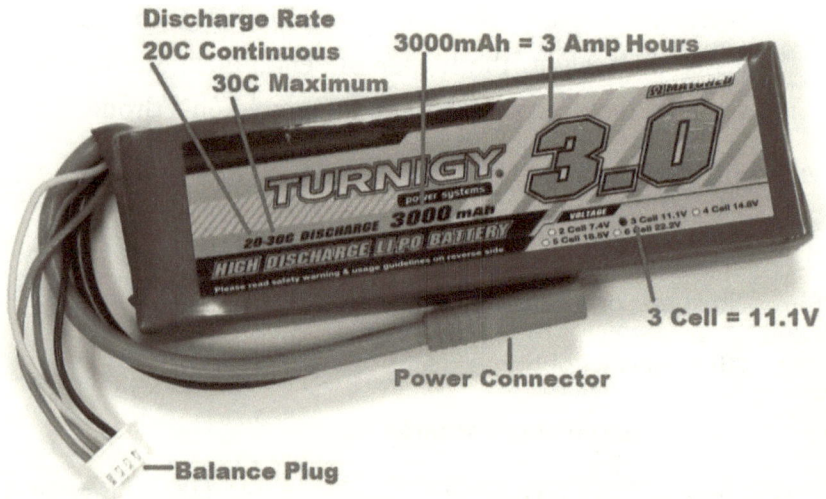

Image 6.11 Battery Parameters (Provided on the Battery Sticker)

The following parameters are identifiable from the product sticker itself on any Lithium Polymer battery.

Capacity: The overall capacity of the battery is defined in terms of ampere-hours (amp-hours or Ah) or milli ampere hours (mAh). Here, the battery capacity is 3000mAh or 3Ah. This defines how much battery can run before the recharge requirement.

On-Board Electrical Components

Discharge Rate: The continuous rate of discharge safe for the battery and motor is 20C (in this example), whereas it has a peak discharge rate of 30C.

This C rating is based on the calculation of the value multiplied by C (capacity of the battery). So, it will be 20*3 (battery capacity in Amps) = 60A. This calculation mainly helps in identifying or selecting the correct motor for your specifications. If it is known that the overall consumption of your motor is less than 60A at any point of time of it in operation, then that motor can be selected for this battery configuration or vice versa.

The number of cells are three here.

Power connectors of different types are available, and it is safe to change the connector by cutting one and soldering the other. However, proper care must be taken while doing this. It is strictly not recommended to cut both positive and negative cable at the same time. This is highly risky and can short circuit the battery.

It is also to be noted that under a certain load, the voltage drop will always take place. The voltage at the idle condition will be different from the load condition. For example, a three-cell battery is fully 100% charged to have a voltage of 12.6V. The moment the motor is powered on with the variation in load there will be a significant drop in the voltage level to a certain extent. In the same way, the moment the drone lands and the motor turns off, the load will be released. It is good to get the voltage calculated instantly after the drone lands, to get the correct, under load voltage.

Ideally, it is good to charge a Lithium Polymer battery to around 95% and discharge it by 70%-80%. The aim should be of landing at least with 20% battery remaining. This totally depends on the type of application for which you are using your drone. Sometimes, you may have to land early; sometimes, it is so far that with 25% return-to-home settings, it may not return and land with the battery remaining. So, it is recommended to set return-to-home as per the site situation and the requirements taking the above-mentioned parameter of 20% into consideration. Otherwise, it may end up crashing your drone or landing it in between its last mission point and home point.

Under any condition, the battery voltage should never ever reach below 3.0V per cell. Going below this limit may damage the cell(s) and, in turn, the overall package will be of no use.

The other most common mistake that people make is reversing polarity connections. This mistake mostly happens while connecting the battery to the electronic speed controller. It is good to check the positive and negative for both the connectors twice before finally making the connection.

b) Lithium-Ion Battery (Li-Ion/LiB)

Rechargeable Lithium-ion batteries have also started getting used for drones and other unmanned vehicle operations, mainly because of a few advantages over their counterpart LiPo.

LiPo	Li-Ion
Stores less power	High power density
Robust and flexible	Rigid due to encased cells
Lower chance of leaking electrolytes	Cells may leak due to damage or other reason
Shorter life span/charge cycles	Larger life span but degrades with time
High manufacturing cost	Low manufacturing cost

Table Comparison Lithium Polymer and Lithium-Ion Batteries[lviii]

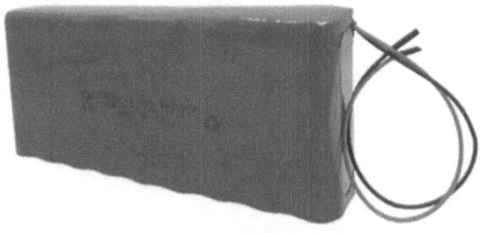

Image 6.12 Lithium-Ion 6S Battery Pack

c) Nickel Metal Hydride Battery (NiMH) and

NiMH batteries are made up of alkaline cells and are capable of providing a capacity of above 2000mAh at very low current requirements.

NiMH batteries are used where the requirement of energy consumption is higher. These batteries are environment-friendly, have a very good life span/charge cycles and are quite safe compared to NiCD batteries. These batteries may or may not have a battery management system for smartly managing the charging and discharging of the battery. It is, of course, good to go for the one with BMS in order to have a better performance.

Typically, these batteries come in four to eight-cell packages of 4.8V to 9.6V with a per-cell voltage of 1.2V.

Parameters	NiMH	Lithium Batteries
Per cell voltage	1.2V	3.6V to 3.7V
Power Density	High	Low
Efficiency	81%	99%
Size	Smaller	Larger
Cost	50% cost of Lithium batteries	High cost

Table Comparison NiMH vs. Lithium Batteries[lix]

Image 6.13   An NIMH Battery Pack

It is mainly used in engine-powered RC aero models where there is the energy required for the operation of few servos and for engine ignition. Apart from using it on non-autonomous manually flown vehicles, these batteries can be used for remote control operations and certain in-house testing.

d)   Nickel-Cadmium Battery (Ni-Cd)

Ni-Cd battery packs offer a very high discharge rate. The life cycle of Ni-Cd batteries is also good compared to other counterparts, whereas their usage in the drone industry is limited to what NiMH batteries are used for. Ni-Cd batteries are not environment-friendly.

e)   Lithium Iron Phosphate (LiFePo4)

Lithium Iron Phosphate batteries are also known as LiFe batteries. These batteries are a pack of cells with one cell comprising 3.2V. The fully charged battery will be of around 3.6V to 3.7V per cell. These batteries are useful for RC hobby-related applications like engine ignitions, etc. These batteries do have a high discharge rating and long life span with around 2000 cycles. However, it has got quite a low energy density compared to Lithium-ion or polymer batteries. Another advantage with LiFe battery is it is lower in weight (which is the most important parameter for being used on a drone). This battery is safer in terms of usage.[lx]

All these batteries can be/are rechargeable in nature but have certain limitations over charge cycles, storage capabilities, decrease in discharging rate, cell bulge, overall endurance/performance decreases, etc. takes place after certain charge cycles. There is no particular number or defined limitation on the charge cycles of a battery, but this definitely depends on the quality of the cells used and battery-built quality. Majorly, it relies on how efficiently it is used during our missions/flying and other testings, along with the manner of maintaining it when we know that it will not to be used for a certain time period.

6.6 Battery Charger

The most important part for the drones is to get their battery charged before its time for the mission. Selecting the charger totally depends on the user as per the user's needs. Chargers of different varieties from single plug-and-play to chargers with multiple-batteries charging-facility are available. Also, the number of cells it can charge up to with fast charging speed (ampere).

If one has multiple batteries with varieties of NiMH, LiPo, Li-Ion, etc., they should select the charger that is capable of charging NiMH, LiPo, Li-Ion and other batteries too. Additionally, it is to be considered that when there are multiple batteries kept on your desk for charging, charging with a single battery charger will slow down your process so, for those with more batteries, you should go for a multiple-batteries charger. These chargers can charge your NiMH, LiPo or any other type simultaneously. It is just about changing certain parameter settings for switching from a LiPo to NiMH for charging your next battery.

Powers: It is important to consider whether the charger is to be used in the field or not. This question mainly comes because the input power source really matters to the charger. Based on the requirements, an AC or DC power source is to be selected. The AC power source will be easier because it will go with any wall socket. To charge a 10000mAh 6S LiPo, the power required is 25.2V*10A = 252 Watts.

Single Port Charger: The single port charger can be used for the batteries with the total number of six-cells (for LiPo). This charger supports Lithium Polymer, Lithium-Ion, NiMH, LiFe, NiCD and Pb batteries. So, these options are switchable and can be adjusted with the help of the given four to five buttons. The charging rate for this charger is 6 Ampere maximum. So, as calculated in the above example, a 6S LiPo requires 252 Watts, whereas as per this charger, it can be charged with a maximum of 6A. The calculation will be 25.2V*6A = 151W. This charger also supports the discharging of the battery in case it is required to be done.

Different charging modes are also available such as:

Storage mode: This is recommended when the battery is not going to be used for some time period. So, to keep it partially charged, this mode should be selected. It will charge the battery to a certain limit of around 60%.

Balance mode: This will charge all the cells in a balanced manner and will charge to the peak of the capacity.

Fast charge: This will charge the battery without balancing and to around 90% of the capacity.

Image 6.14 A Plug-and-Play Charger for 1, 2 and 3 cell LiPo Batteries

Multiple Port Charger: It can be used for batteries with the total number of six-cells (for LiPo). This charger supports Lithium Polymer, Lithium-Ion, NiMH, LiFe, and NiCD batteries. This charger also supports the discharging of the battery in case it is required to be done.

Image 6.15 A Multipurpose Charger for Batteries like LiPo, Li-Ion, LiFe, NiMH, NiCd, etc.[lxi]

CHARGER SPECIFICATIONS					
Battery Type	Number of Cells	Voltage Level	Max. Charge Vol.	Max. Fast Charge Rate	Min. Discharge Vol.
LiPo	1 – 6	3.7V/Cell	4.2V/Cell	1C	3.0V/Cell
Li-Ion	1 – 6	3.6V/Cell	4.1V/Cell	1C	2.5V/Cell
NiMH	1 – 15	1.2V/Cell	1.5V/Cell	1C – 2C	1.0V/Cell
NiCd	1 – 15	1.2V/Cell	1.5V/Cell	1C – 2C	0.85V/Cell
LiFe	1 – 6	3.3V/Cell	3.6V/Cell	4C	2.0V/Cell

Table Charger specifications[lxii]

It is always recommended to charge the batteries with balance charging. This will charge all the cells equally and prevent overcharging of any cell.

Most of the chargers come with different plugs to connect to the batteries. If it is not fulfilling your requirements, it is easy and good to change the connector or have a conversion cable separately.

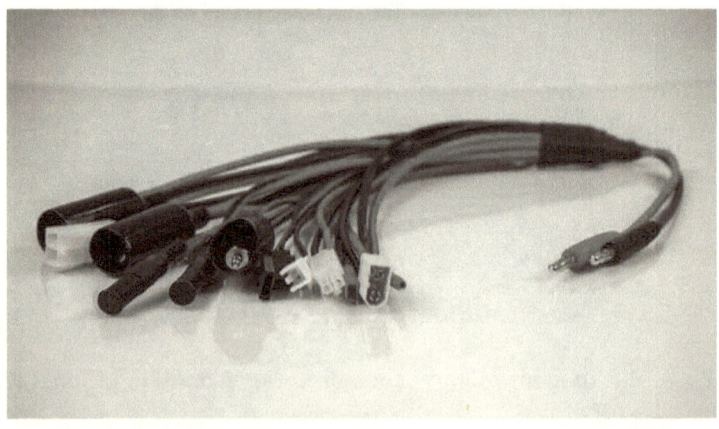

Image 6.16 Various Connectors for Charging Varieties of Battery

Overall Instructions

- Charging Steps: Turn on the charger. Setup the parameters you want to charge the battery for. Connect the charging cable to the charger. Connect the balancing plug of the battery with the charger. Connect the main battery plug with the charging cable and start charging.
- Do not leave the battery unattended and on overnight charging.
- Handle with care while transporting.
- Do not overcharge the batteries.
- If it is not in use, do charge the battery under storage mode.

Chapter Insight

The selection of servo, electric motors, battery and relevant electronics for a drone matters a lot as the efficiency of a drone is dependent on the propulsion system and its settings.

CHAPTER 7
UAV PAYLOADS

Payloads are loads that can be carried by a drone in order to accomplish certain missions and applications with it. These payloads are normally in the form of cameras, explosive warheads, logistics (parcels), certain sensors like LIDAR, robotic arms, etc. As per the trending applications in commercial and military services, our main focus here will be cameras and LIDAR sensors. These payloads are very important for survey and surveillance applications. The description provided for any particular model under any type of camera is only for the basic explanation of different camera types and their applications. There are other different models and cameras available for selection based on our purposes.

7.1 RGB Camera

RGB Cameras are regular cameras with red, green and blue colour models.

1. **Sony a6000**

    Sony a6000 is Sony's new mirror-less camera that has got certain additional features along with other common features of Nex series cameras.

This RGB camera is quite preferred for a drone-based survey when it comes to self-developed or a customised drone as it has got the fastest AF which captures sharp images and is capable of capturing the images remotely which is an important requirement for a drone user.

Image 7.1   Sony Alpha 6000

Sony a6000 can be controlled remotely from the ground control station for camera triggering, which is done through the autopilot system.

Key Features

- 24.3 megapixels APS-C CMOS sensor
- Bionz X image processor
- Hybrid AF system with 25 contrast-detect and 179 phase-detect points
- Built-in flash + multi-interface shoe
- 11 fps continuous shooting with subject-tracking
- 3-inch tilting LCD with 921,600 dots

- OLED electronic viewfinder with 1.44M dots
- Diffraction correction, area-specific noise reduction and detail-reproduction technology
- Full HD video recording at 1080/60p and 24p; clean HDMI output
- Wi-Fi with NFC capability and downloadable apps
- Image size: large - 6000 x 3376
- Weight: 350grams
- Dimensions:127 x 67 x 45mm
- Not waterproof

Drone-based Applications

- Land mapping
- Forest area survey
- Highway surveys
- Any large area mapping

7.2 Thermal Camera

Thermal cameras are the temperature detecting infrared cameras which mainly capture and allow us to identify the temperature of multiple objects in a single picture. All the objects emit infrared energy according to their thermal properties.

1. **FLIR Duo Pro R**

    The new FLIR Duo® Pro R combines a high resolution, radiometric thermal imager, 4K colour camera and a full suite of on-board sensors to bring you the most powerful dual-sensor imaging solution in the world for small commercial drones.[lxiii]

Image 7.2   FLIR Duo Pro R Camera

Drone-based Applications

- Solar utility inspection
- Power line monitoring
- Tower inspection
- Structure inspection
- Inspection of pipelines
- Fire fighting and rescue operations

7.3 Multispectral Camera

Multispectral cameras capture images at different wavelengths of light like the near-infrared, red band, red edge, blue band, green band, etc. with a precise set of light wavelengths for each. Their application is in precision agriculture.

Mica Sense Red Edge-MX

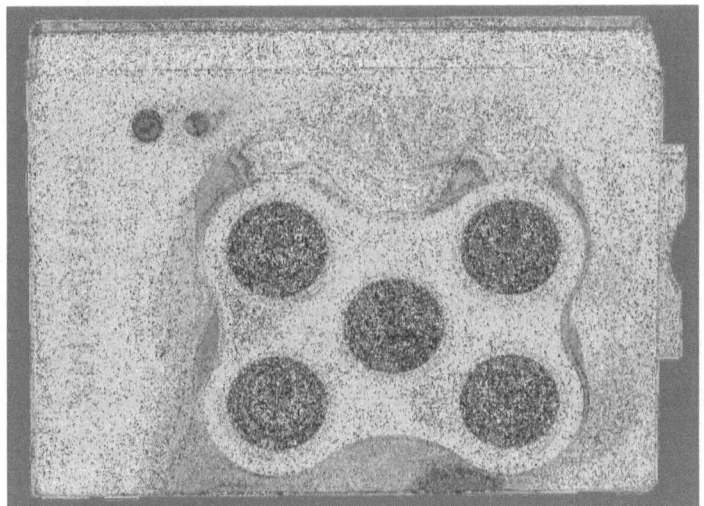

Image 7.3 Multispectral Camera from Mica Sense

Key Specifications[lxiv]

- Spectral bands: blue, green, red, red edge, near-infrared
- Wavelength: blue (475 nm centre, 32 nm bandwidth), green (560 nm centre, 27 nm bandwidth), red (668 nm centre, 14 nm bandwidth), red edge (717 nm centre, 12 nm bandwidth), near-infrared (842 nm centre, 57 nm bandwidth)
- GSD: 8 cm per pixel (per band) at 120 m (~400 ft) AGL
- Field of view: 47.2° HFOV
- Weight: 232 grams

Drone-based Applications

- Monitor the evolution of crops
- Track progress of seeding efforts

- Disease identification
- Fertiliser management
- Crop health mapping

7.4 LIDAR Sensor

LIDAR is a "Light Detection and Ranging" sensor that creates a 3D model of the surveyed area. The work of this sensor is based on the emission of LASER light and the pulses that return to the sensor. The time taken by the pulses to return is calculated and the variation in the distance is measured.

Drone-based LIDARs weigh between 500 grams to 1 kg based on the specifications. Most of the LIDARs have horizontal FOV of 360°.

Image 7.4   Velodyne LIDAR

Drone-based Applications

- 3D Point cloud generation of an area
- Forest mapping
- Terrain mapping
- SLAM for mining

7.5 GIMBAL

A gimbal is a support as well as a camera mounting platform which allows the rotation of an object of about two or three axes. For drone-based photography, it is recommended and good to have the camera mounted on gimbal as it provides stabilisation in the air in terms of stability in the wind, vibrations dampening and allowing the camera to remain stable at one position.

Image 7.5 Gremsy Gimbal for Sony Alpha 6000 Camera

Gimbals are powered by two and three brushless motors for two and three axes of rotations respectively. These three axes are pan (yaw), roll and tilt (pitch). The operation of rotation can be done with the help of remote control or ground control station. It consists of a Gimbal Control Unit (GCU) which controls the motor operation as per the input received for the angle change.

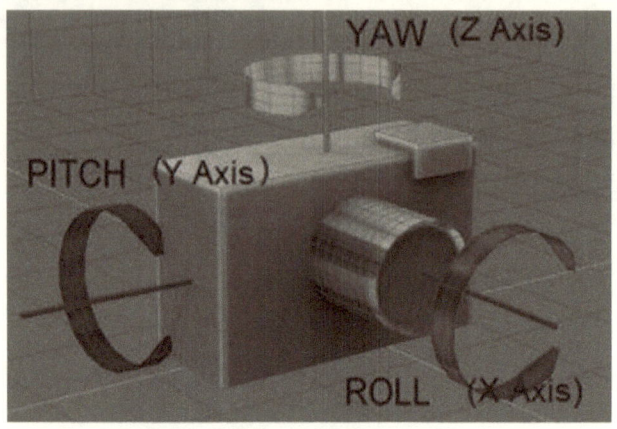

Image 7.6   Roll, Pitch and Yaw Axis Rotation of a Gimbal

7.6 Other Payloads

Any designated payload can be added to the drone as per the requirements. Other payloads are EO/IR sensors, sprayers for spraying drones, robotic arms, logistic loads, any other cameras, weapons, SAR, etc.

Some high-end drone manufacturers use a dedicated depth camera to construct a 3D model using stereoscopic techniques.

It is to be noted that certain COTS drones come with a pre-added camera of the seller's choice. So, it is necessary to analyse the camera type and its specifications, and then finalise it for purchase and your use.

Chapter Insight

A sensor in the form of a payload captures images, videos, or points based on the sensor used. This collected data is analysed using certain software based on the client and the user's requirement, which is considered as data processing. It is not necessary for a drone to have the above-mentioned sensors (only); it can be anything else, as new sensors are coming in the market in a very short period of time now. Even a parcel can be considered a payload.

CHAPTER 8

Remote Controller and Communications

8.1 Remote Controller

Remote controllers are electronic devices which are used to operate a UAV from the ground through some communication medium. The control operations are done by the pilot, as per the need of flying. A remote control system has two modules; one is a transmitter which is used to control the vehicle movements and another is a receiver which receives the next operation commands from the remote control (indirectly, from the person using the remote control). A remote control transmits the commanding signal to the receiver placed on the drone. The signal received by the receiver is then sent to the flight controller in order to control the drone during flight. For an RC plane, without an autopilot system, the command received by the receiver will be converted and forwarded for different servo or motor movements.

The basic remote control transmitter will have four sticks/joysticks. The image shown below is of a typical four-channel remote controller.

Image 8.1   A Four-channel Transmitter[lxv]

The four channels are for four different controls on your drone. These four controls are Throttle (Thrust), Aileron (Rolling moment), Elevator (Pitch moment) and Rudder (Yawing moment). These sticks are gimbal sticks and move up and down as well as right and left at any point of the angle. These sticks are sensitive. It has got springs so the moment it is left from any position, it comes back to the centre. So, for a constant roll, pitch or yaw angle requirement, one will need to control the stick continuously for a particular time period until the desired angle or turn is achieved on the aircraft. The throttle stick on most of the RCs is manual, without spring, and remains at the position where you leave it as the power required for the motion of an aircraft should be constant or as per the requirement, but it can't be always neutral or 50% with a stick in the centre position.

The above shown was the most basic model with no extra control switches available. There is no display screen available on which we can look and make changes in the parameter settings as per the need. So, there are other advanced transmitters available with the display screen where we can check and do all the parameter settings. There are also remote controls available with the in built screen. This is quite useful in certain applications for flying as it eliminates the need for an external screen for live data feeds, mission planning and other settings.

Image 8.2   Taranis FRSKY Remote Controller with a Receiver

Most of these remote controllers are compatible with planes (glider, aerobatic), helicopters, whereas some are compatible with all types of planes, helicopters and Multirotor. Although most of the latest remote controllers are available with all types of vehicle controls, it is best to crosscheck before selecting one as your requirement.

Image 8.3   A comfortable display for change in settings[lxvi]

As shown in the above image, the model selected for the setting is a multirotor. Under the multirotor category, the configuration type is to be selected.

Also, there are different modes of RC available for purchase and use. These modes are just changes in the stick operation that what stick can be used for what purpose. The image defining the same is shown below.

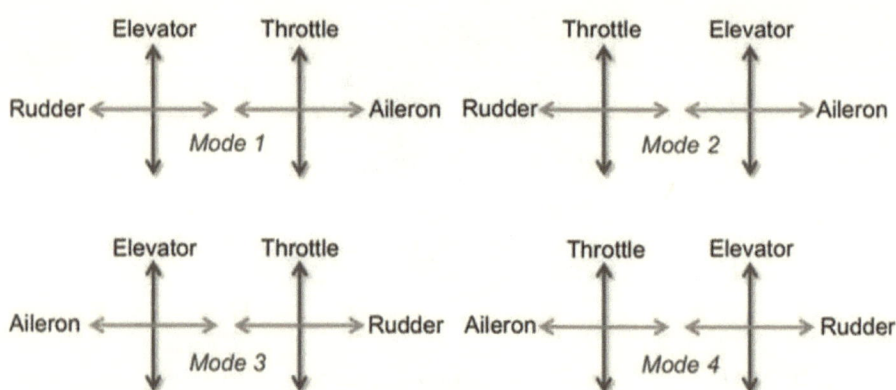

Image 8.4 Stick Defining for Mode – 1, 2, 3, and 4 Types of Remote Controllers

Four modes are shown in the above image.

In mode 1 and mode 3 remote controllers, the throttle remains on the right-hand side. The only change we can see is that the rudder and aileron have been replaced by each other in mode 3 compared to mode 1. Mode 1 was normally available till 2015.

In another set of mode 2 and mode 4, the throttle is on the left-hand side, whereas the rudder and aileron replace each other in mode 4 compared to mode 2.

Typically, in an RC, there are gimbal sticks for the above shown four controls. Apart from that, there are other switches available. These switches can be two-way/three-way toggle switches and knobs. It can also just be a trigger switch. These switches can be used for different features like gimbal movements, camera triggering, flight mode changes, etc.

The receiver may come with a remote control purchase in the bundle of combos or else. It is necessary to go through the remote controller manual for the description of the list of receivers the RC is compatible to fly with. There are necessary steps to be followed for the one time setting of a new model on the transmitter and assigning a receiver by following the binding process. So, in case you are trying to connect your RC to the drone for the first time, it is important not to forget to bind it.

This is to note that there is a difference between RC flying and drone flying. It is an RC flying when an aero model or plane flies without any autopilot assistance and in manual mode. The vehicle during the whole flight is controlled by the pilot using a remote controller. Under such conditions, it is necessary to have an idea of the range limitation for the communication between the RC transmitter and the receiver. Because once the vehicle goes out of transmitter range and it is not coming towards you, it may turn out to be a really bad day for you.

These remote controllers communicate with the receiver at a particular frequency for sending commands for further operations. The most common/popular frequencies these days are 2.4 GHz, as it ensures a solid link between the drone and RC even when there are multiple drones operating at the same frequency. This has replaced the traditional crystal technology of 36MHz, 40MHz, etc., systems which offered the longer wavelength, yet dramatically leading to the interference and connecting with different models during multiple aircraft operations.

With 2.4GHz, the reliability increases with the shorter wavelength which decreases the signal penetrating power-range decreases. However, this can be resolved with a higher power output of radio-frequency in built in the radio. There are also external modules available that are frequency range boosters.

The method of communication between a transmitter and a receiver is based on a protocol. This includes digital protocols SBUS and DSM2x. The older protocols were pulse position modulation, pulse width modulation, etc. The trending SBUS protocol transmits control signals from receivers to servos as well as flight control systems. The response time is faster in SBUS as compared to PPM and PWM, as PPM and PWM have delays of around 50ms.

Among the currently available transmitters, some are OpenTx. OpenTx transmitters are reprogrammable. Different functions and features can be added as per the need of the user. High-end remote controls are not always OpenTx and are with limited preloaded functionalities. For normal users, even these functionalities are not fully utilised by them.

The low-end RCs of four to six channels do not have the capabilities of saving different models inside, whereas the high-end RCs have enough inbuilt storage that it can store some 30 models. With an SD card, this can be expanded further to 100 models. So, it is really like 100 vehicles on one transmitter.

8.2 Receiver

A receiver is a module used for communication with the transmitter. This receiver has two antennas. These antennas are recommended to be placed 90° angle from each other. The receiver has around eight vertical three-pin rows. These pins are used for different crimp connections for servo, motor or any other system operation. The two horizontal crimps are used for battery and SBUS/PPM connection with the autopilot system.

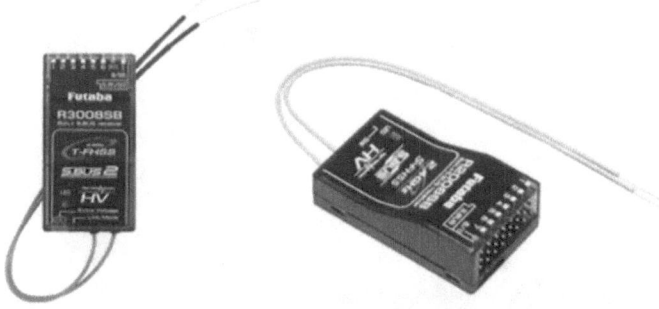

Image 8.5   Receiver of a Futaba Transmitter

An SBUS receiver that connects to the flight controller system and thereby is connected to other devices for control and operations.

8.3 Telemetry System

The telemetry system is the part of communication between two devices that are connected wirelessly and thereby transferring the data. In the field of drones, it works the same way. It provides us with the details of where the drone is flying and many other details. Telemetry systems use certain frequency bands in order to communicate and transfer the data/receive the information. Most of the available telemetry systems come with a frequency of 433MHz, 915MHz, 2.4GHz, etc. and their range is totally dependent on the frequency of telemetry, transmission power, weather, location, obstacles, etc. However, this needs to be noted that not all the frequency bands are delicensed in India. So, it becomes necessary to check for compatibility and compliance with Indian standards before purchasing one for you.

Image 8.6 A 915MHz 3DR Telemetry System

Telemetry systems may not be required many a time, but for all those using a ground control system with a Pixhawk/ArduPilot series autopilot system, it becomes necessary to have one as it becomes the only source of communication from your drone to the ground control station device available in your hand. The above-shown telemetry is compatible with Pixhawk series autopilot systems. There is a simple cable connection required for air module connecting to Pixhawk and ground module connects with your computers/tablets/phones. Once there is communication, updated parameter details will show up in GCS. (Refer to chapter 10: Ground Control Station System.)

Chapter Insight

Remote controller these days is mostly used for practising and testing purposes. Once the mission mode is ON, the flight controller carries out the mission without any continuous need of pilot input.

CHAPTER 9
Autopilot System and Sensors

9.1 Autopilot System

An autopilot system is the heart of a drone that has the capability to control all the moments, motion and operations automatically, when it has got a pre-assigned parameter and a lot more functions to carry out. The selection of an autopilot system is something that is very much important and quite crucial too, at times.

A flight controller system is a control system that manages the drone flight. Like, it is necessary to control the power and RPM in a Multirotor flight. This depends on the command received from the pilot (through remote controller) and even on the internal self-management (when the pilot is not giving any control input). Apart from controlling just powers and RPM management, there are sensors available on board which gives information about the dynamics and position of the system.

Users these days consider flight controllers as autopilot, and vice versa. So, it is quite important to clear out the difference.

An autopilot system is a complex system that manages and controls the operation of drone flying in certain different ways with much wider parameters compared to a flight controller. Apart from simply managing drone flying, it also does positioning. There are sensors on board for altitude holding, flying, controlling, carrying out autonomous flying once the path is provided, and clicking the images on the path autonomously. There are different modes of flying available to be setup before the flight. It also has the return-to-home facility under which the drone will return to its home location in occurrence of any problem. Whereas, a flight controller system may not be able to go for an autonomous flight but will be able to keep the drone stable when flown the drone in manual mode. So, if we do not go by the name, the autopilot system can be called as an advanced flight controller.

There are also a few open source flight controllers available which can be used on our vehicles. These advanced flight controllers offer much more operations and keep getting updated for the latest version, being an open source. For example, a Pixhawk series of autopilot systems can be used for any Multirotor frames. The same autopilot also allows us to configure it for a Fixed-wing vehicle, a rover, a boat, a submarine/underwater vehicle and a helicopter.

Image 9.1   Pixhawk 2 Cube Flight Controller

After the configuration of different vehicles, it supports different varieties of sub-vehicles. In the case of Multirotor, it supports Quadcopter, Hexacopter, Octocopter, Tricopter, Y6Copter, X8Copter, etc. The same goes for Fixed-wing vehicles, where it supports normal plane configuration, V-tail, A-tail, delta wing, etc. This way, Pixhawk becomes a versatile system for all unmanned vehicles.

Autopilot System and Sensors

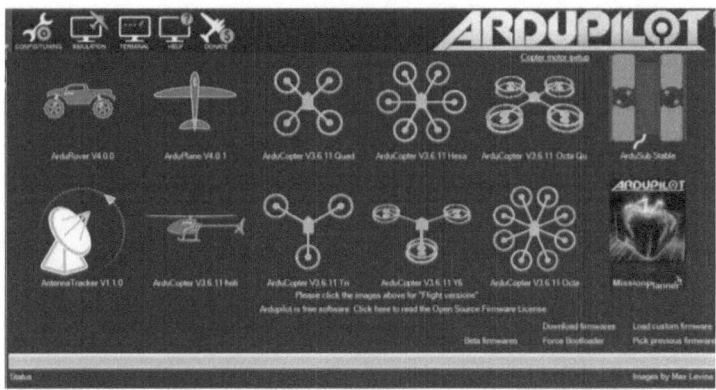

Image 9.2 Screenshot of Mission Planner Vehicle Selection Configuration

All these systems get the facility of auto take-off, auto land or return-to-home in case of any issues with the drone or when the battery gets lower than the preset parameter. Transmit their live location, execute preloaded commands, carry out missions, and many more things that can be explored during practical usage.

Throughout the mission, it is easy to identify the parameters like distance from home location, flying altitude, speed, climbing speed, vertical speed, location of the vehicle, battery remaining and much more. The things mentioned here are the basic settings and basic parameters that are available on HUD during the mission.

Image 9.3 Mission Planner GCS Main Screen

Flight modes supported by this system include around 25 different modes which include stabilise, auto, position hold, manual, acro, etc.

Carrying out different types of missions is clearly possible with the Pixhawk autopilot system and other autopilots as it provides planning in grid mission, double grid, manual waypoints, etc.

9.2 Internal Sensors (in Autopilot System)

1. **IMU**: IMU stands for Inertial Measurement Unit. It is a combination unit of different sensors. A device that contains one or more accelerometers and gyroscopes in order to measure the acceleration force exerted upon the sensor and orientation as well as the angular velocity, respectively. This may also include a magnetometer. This set of all three sensors are in the entire three principal axes, pitch roll and yaw.

2. **Accelerometer**: It is a sensor used to measure the proper acceleration force exerted. It is an electromechanical device that senses the static/dynamic force of acceleration. This is dependent on vibrations and displacement. This can weigh from 1 gram to 250 grams. The smaller, the more detailed the data will be. A 250 gram accelerometer will be used mostly for rockets.

3. **Gyroscope**: A sensor used for the measurement and maintenance of orientation and angular velocity of a vehicle. A gyro wheel which is attached to three endpoint gimbals in such a way that it allows the wheel to rotate. There are three such wheels for each axe of rotation. It is measured in revolution per second. Gyroscopes are the size of 1 micrometre to 100 micrometres, and that is due to MEMS. A gyroscope GIF will help you understand this in a better way.

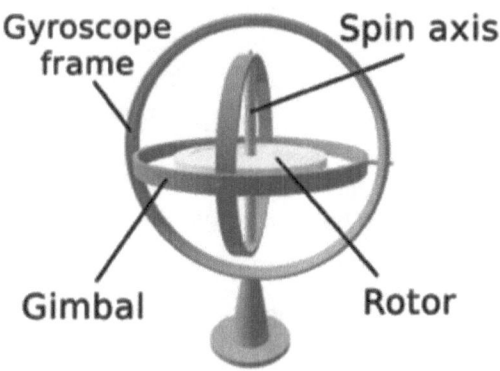

Image 9.4 Gyroscope

4. **Barometre**: It is a pressure sensor, used to measure the altitude level of a vehicle based on air pressure. It is similar to pressure altimeter, but the uses are different.

5. **Magnetometre**: It is used for the heading reference. The heading reference information cannot be achieved from the accelerometer or the gyroscope. So, the magnetometer is required to know the facing position of the vehicle. Sometimes, this is the part of an IMU.

6. **GPS Receiver**: GPS (Global Positioning System) technology is a satellite navigation system used to determine the position of an object. It is used to identify the location of the vehicle. A GPS receiver works on GPS technology where there are 24 satellites in space orbiting around the Earth. These GPS satellites are in synchronisation with the nearby four satellites in their line of sight. The GPS satellite broadcasts messages that include information about its current position, orbit and time stamp of that position. The GPS receiver then combines such information received from multiple satellites and calculates it using a triangulation method. For this purpose, a minimum of four satellites is a must. Mostly for drones, we never

take-off before getting 12 to 15 satellite connections received by the GPS.[lxvii]

A GPS receiver is connected with the autopilot system of the drone. This then allows us to check the number of satellites before taking off the drone. Once we get enough satellites, the home point gets recorded, which is really important as we always want our drone to return to the home location when the mission is done or in case some issues occur.

An autopilot system may allow external sensors for the operation through its software and hardware possible configurations. This totally depends on the autopilot system purchased for your usage. There are many external sensors that are compatible and purposeful for many applications. Some of the important sensors that can be configured with an autopilot system are as follows:

9.3 External Sensors

1. **Airspeed Sensor**: It is a pressure sensor that measures the speed of air passing through the aircraft. It compares static and dynamic air pressure with the help of pitot tubes.

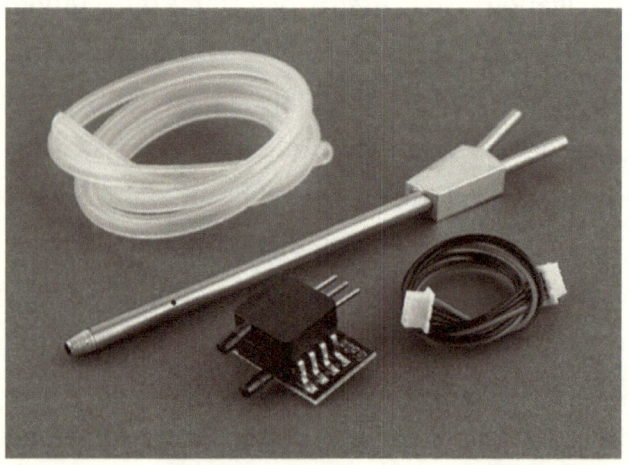

Image 9.5 Analog Air Speed Sensor (Google images)

2. **Obstacle Avoidance Sensor**: An Unmanned Aerial Vehicle flying on the prescribed path at any certain altitude may not be able to detect any obstacle on its own. For example, a drone flying at an altitude of 30metres can hit a building or may be a transmission tower. To detect and avoid it are important tasks as without such sensors, we may have certain broken parts of drones in our hand. So, for the safety of the drone, it is good to have an obstacle avoidance sensor. These sensors can be in any variety of forms with different range specifications. Let us have a look at the obstacle sensors that are commonly used for drones.

3. **Ultrasonic Sensor**: It is a device that transmits ultrasonic waves which are received after colliding with the obstacles, if any. With the time of reception of the signals, it identifies the distance to the obstacle from the drone.

Image 9.6   An Ultrasonic Sensor by APG Sensors[lxviii]

Above-shown is a long range ultrasonic sensor which is capable of detecting objects up to 50 feet. Most low-category ultrasonic sensors are not capable of detecting obstacles beyond 30cm.

4. **Range Finder**: It is a sensor used in the drone to identify the distance to the ground. It is attached to the bottom part of the drone so that it becomes easier to identify how above the drone is

flying from the ground. Note that this is only used when there is no barometre sensor used. LIDAR range finders are quite common for such purposes.

Image 9.7   LIDAR Lite V3 from Garmin

The above image is of a LIDAR Lite V3 optical distance measurement sensor from Garmin which transmits an invisible beam of laser from one end and receives from the other end. With the reception time, the distance is identified. The sensor has the capability to identify objects from a 130foot distance.[lxix]

Additionally, there are certain configurations available for payload release, movement of the camera through gimbal and triggering of the camera.

Apart from Pixhawk series and DJI flight controller boards, there are many other closed source flight controllers available in the market like Eagle Tree Systems, Micro Pilot, etc. which can perform precise missions in more or less the same manner.

Chapter Insight

A flight controller is the brain of any drone. The more accurately it controls the drone, the more stable and smooth flying the drone will have. Precise integration, calibration and tuning are necessary to have a smooth flight.

CHAPTER 10
Ground Control Station System

10.1 Introduction

A drone's operation can be planned, executed, controlled and monitored using a ground control system that is connected with the vehicle's autopilot system. The same ground control system is also responsible for drone system setup and related system integrations. An autopilot system should be mounted on the drone in such a way that the forward arrow of the autopilot matches with the front of the vehicle. Many a time, it is required to keep the autopilot offset, not in a manner that it can point towards the forward of the vehicle. This is allowed in many advanced flight controller systems where the offset angle is to be set in the parameter configuration so that it can nullify the offset and controls can function properly.

All these settings, planning, execution and monitoring vary with the changes in the autopilot system development. So, different products will have different settings and options available for carrying out the operations. So, here, we will discuss the major parameters that we can get over a ground control station. These parameters will be in miscellaneous terms and are explained collectively in a common format. It is possible that any of the below explained parameter settings or configuration may not be available in the ground station software that you are using or are planning to use. It is also recommended that for a drone, types of missions to be carried out should be decided before purchasing a flight controller board/autopilot system so that we do not miss out on the operation that we want to perform by selecting the right brain for our drone. The illustration images shown here for an explanation might be of some particular type of software system but the main concept behind all these software remains the same.

A ground control station communicates with the drone wirelessly using a telemetry system that needs to be linked up with the GCS at one end and with the autopilot of the drone on the other. Once the correct connection is established, it will start showing the updated parameters of the vehicle every moment. There are multiple parameters updated; however, not all can be listed. Some of the major parameters are satellite connections, telemetry signal strength, flying altitude, moments of the vehicle, flying speed, distance to the home location, showing the current location of flying on a map, battery usage, any errors, etc.

10.2 GCS Head-up Display Data

The overall main screen of the mission planner application allows us to connect or disconnect the ground control station with the drone. This connection is possible with the help of a telemetry system connected with the autopilot system. Once the drone and the GCS are connected, the current drone status and parameters are received on the main screen of GCS.

Ground Control Station System

Image 10.1   Mission Planner Application – Flight Data Screen

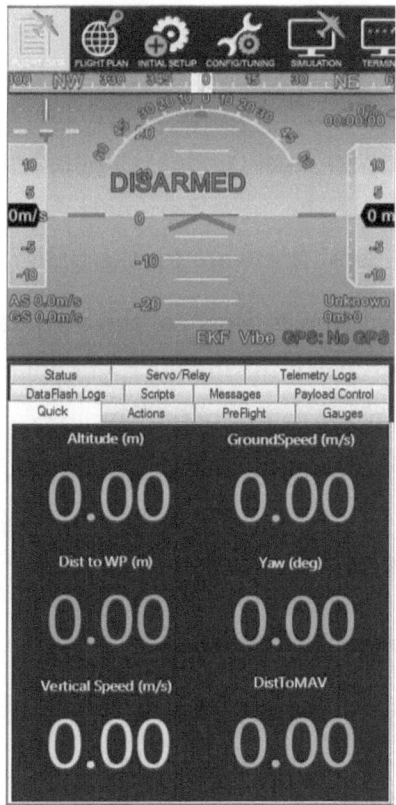

Image 10.2 HUD and Quick Screen that Provides Selected Parameter Information During the Flight

Throughout the mission, it is easy to identify the parameters like distance from home location, flying altitude, speed, climbing speed, vertical speed, location of the vehicle, battery remaining and much more.

The above image describes different parameters that can be monitored during a flight on the flight data screen. Moreover, GCS provides an Action tab from where flight modes can be changed without the use of a remote controller.

10.3 Vehicle Parameter Settings

Almost all the controller/ground stations provide one-time system settings where the type of vehicle is defined like Multirotor and if it is a Multirotor, it is a Quadcopter. If it is a Quadcopter, it is an X-configuration Quadcopter. With this selection of a particular model of the vehicle, the firmware for that particular model is uploaded to the flight controller. This is important as the next parameter settings of the vehicle are dependent on the right firmware for the right drone type.

Once the above step is finished, the next step comes is to calibrate the sensors mounted on or inside the flight controller. Sensor calibration includes accelerometer calibration, compass calibration, magnetometer calibration, radio calibration and ESC calibration in the primary setup.

10.4 Calibrations

1. Accelerometer Calibration

In this calibration, the levels of the vehicle are calibrated for all sides. The accelerometer calibration is performed by putting the drone on a flat levelled surface, then turning to the left and right side of the drone, nose up, nose down and the upside-down position. This will be informed as a step while performing the calibration.

The turning of the drone to the left and right should always be considered from the rear part of any drone.

2. Compass Calibration

In this calibration, the Earth's magnetic field intensity is detected. Due to certain magnetic interferences, it may get disturbed or can show the wrong heading of the compass. To keep it in the correct position, it is required to calibrate the compass. For compass calibration, the vehicle needs to be rotated in three axes.

3. Radio Calibration

The transmitter that we are going to use for the operation of the drone is to be calibrated with the ground station. This is for the calibration of the switches that are going to be used during operation (assigned switches), though it is always good to calibrate all the switches in one go. Therefore, when any unused switch is assigned to some new sensor, at that time, calibration may not be required.

During the radio calibration, the transmitter sticks are to be moved up and down, toggles and knobs, everything. The parameter value will show up as per the changes happening in the PWM values.

In case additional sensors are used, the calibrations for those sensors or electronics should be carried out.

10.5 Mission Planning

Mission planning is something that allows you to create a flight plan so that the drone can fly autonomously on the provided trajectory. Missions can be planned in different formats like way points, rally points, grid missions and terrain following. These missions can be planned from the office and can be saved for future executions. This can save more time and effort during field executions.

It is important to check and set home points when not on the ground. Basically, when the drone gets armed, the home point is set. With the home point set, the vehicle can return to its home location when RTL is executed.

1. **Waypoint Mission**

   For a multi-waypoint missions, the points where we want the vehicle to go can be set. Multiple waypoints can be added by just clicking on the desired locations on the map. Altitude can be set differently for each of the waypoints.

   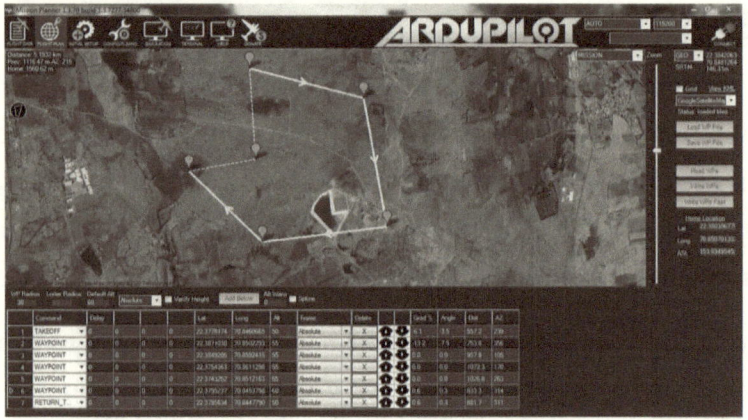

   Image 10.3   Example of Waypoint Adding Up in Mission Planner

   In the waypoint mission, as shown in the above image, three waypoints are added. More waypoints can be added. At the bottom, a list of waypoints is provided from where the order or the points can be changed. Other parameters can be selected for take-off, landing or return to launch command directly so that they can be executed autonomously.

2. **Auto Grid Mission**

   In an auto grid mission, the grid area of interest for flying drones is provided, and the vehicle carries out the mission in a back-and-forth motion in "lawnmower" pattern over the selected area. This grid can be created by selecting an area directly and setting other flying parameters and image capturing settings. Also, such grid missions can be planned by uploading a KML and then setting parameters for flight and photogrammetry.

Image 10.4 Auto Grid Mission in Mission Planner

On other applications like Drone Deploy, the auto mission grids for 2D Photogrammetry can be planned for the selected area by uploading a KML of the area. Certain parameters need to be set for image capturing and flying altitude of the drone. Once the parameters are set, the application directly creates a grid mission for the provided area.

Image 10.5 A Planned Mission over a Lake

Generated grid lines are also to be considered as a flight path for the drone. The given flight path can be changed by changing the angle of the grids. These grid missions can be converted into double grid missions where the missions will be carried out autonomously in horizontal and vertical grids. This is basically required when one needs to carry our 3D Photogrammetry of the area of interest.

Image 10.6 A Double Grid Planned Mission over a Portion of a Lake

The planned mission may not be finished in a single flight. Under such conditions, the drone can be called return-to-home location or it would autonomously return-to-home after calculating the battery remaining for carrying out the mission. It can resume from the same grid line from where it was left on the previous flight after the battery is changed.

These missions are aimed at all types of aerial drones, VTOL, Fixed-wing and Multirotor systems with a defined autopilot system.

Certain parameter settings may be required while configuring the control system of different vehicles like VTOL, V-Tail Fixed-wing, etc.

3. Terrain Following

At times, the planned mission has quite a hilly terrain where the elevation variation may go above a certain limit. If it is sloppy, the quality of the captured data will be affected. If it is hilly, there is a chance that the drone may crash. For that purpose, terrain-following features are quite useful as the flying altitude remains the same from the ground at every location whether there is elevation level increase or decrease. Of course, planning is required to be done beforehand in the recommended format; however, it still pays off the efforts of developing the drone by saving it from a crash.

10.6 Flight Modes

The description provided so far in the ground control station is related to mission planning and flight planning. However, those flights are to be carried out in different modes as per the mission requirement. Different types of flight modes that are commonly available are positioning, alt-hold, auto, stabilise, manual, loiter, attitude mode, return-to-land, etc. These flight modes fulfil some drone flying and controlling requirements. These include even a mode that can be used for flight training and a mode that can carry out a full mission. Note that it is not always necessary to fly in mission/automatic flight mode. Manual flights can be carried out without any need for mission planning for testing and training purposes. Additionally, these modes are not a fixed mode and not necessary to be decided before take-off. These modes are switchable to another mode based on the modes assigned to the switch. However, it is advised to keep the required modes set to a dedicated switch so that while flying, you do not miss out on a mode. Let us check one by one what they really used for in your mission.

1. **PIXHAWK Autopilot Flight Modes (for Planes and Multirotors)**

    The Pixhawk series flight controller provides many options when it comes to the mode of flying being a versatile flight controller. These modes are quite important and appropriate as per the flying needs. The modes that are required for basic flying are explained here.

a) Auto Mode

A drone, when provided with a flight path autonomously, will follow the path to execute the mission, without any intervention or input controls provided by the pilot. This flight mode can be executed before take-off as well as in the air.

The path is followed autonomously, and when the mission points or grids are over, it comes back to the home location under return-to-land mode. A copter can land automatically if the landing command is provided.

For example, there is a square path prepared on a site as a pre-planned mission with four corner waypoints. Once the drone takesoff, it starts following these waypoints in serial numberings. Once the flying on the provided path is done, it will activate the return-to-land mode.

b) Manual Mode

The manual mode mainly refers to piloting mode. The input provided by the pilot will be the output function of the vehicle. So, with the right roll input, the drone should go towards its right side. It holds the altitude and takes the reference of GPS for positioning.

c) Stabilise Mode

Switching over to stabilise mode will automatically maintain the vehicle level. However, the vehicle can be flown manually under this mode. This is a kind of semi-manual mode where all the control inputs are to be given by the pilot but after the movement of the vehicle, it gets self-levelled. In this mode, roll, pitch, yaw and throttle, all four controls can be manually operated.

This mode is not recommended for learners. In the case of autopilot failure, the flight mode should be switched to stabilise mode so that at least the level of the craft can be maintained and land safely.[lxx]

d) Loiter Mode

Loiter mode maintains craft stability in terms of maintaining altitude, holding position and constant heading. It is similar to the manual mode, but it needs fewer control inputs as it manages itself in the air when the controls are released. This mode is recommended for learners.

e) Altitude Hold Mode

As the name suggests, the altitude remains on hold and in control. So, if the drone is flying at 50 metres and switched to Alt-Hold from any other mode, it will maintain that altitude. However, the vehicle will drift due to air pressure/wind. So, there is no positioning control under this mode. In this mode, the roll, pitch and yaw can be controlled manually. This mode is recommended to learners.[lxxi]

f) Return-To-Location Mode

Once RTL mode is activated, the drone will return to its home location, i.e., the point that was recorded before take-off. It is necessary to set the altitude of the RTL mode as per the requirement so that the vehicle can fly safely and return without any obstacle issues. Under the auto mission grid, corridor and waypoint missions, the drone will automatically return to the launch location. There are certain supportive parameters that may need to be set before initiating the mission, like loitering time at RTL, landing speed, flying speed while returning to land, etc.[lxxii]

2. **DJI Drones/Flight Controller Flight Modes (for Copters)**

   a) P-Mode (Positioning): P-Mode works best when the GPS signal is strong. The aircraft operates, stabilises and navigates between obstacles based on GPS reference and vision sensors. This mode is used for carrying out the missions autonomously. Manual flying can also be carried out under this mode.

   b) S-Mode (Sport Mode): For this mode, the gain values of the vehicle are to be adjusted for enhanced manoeuvrability. The maximum flying speed increases to 20m/s. However, it is necessary to note that the obstacle avoidance sensors are disabled and the vehicle will not be able to automatically avoid obstacles in its flight path.

   c) A-Mode (Attitude Mode): In this mode, neither GPS nor the obstacle avoidance system is available. The aircraft controls the altitude using a barometre. Under this mode, the drone will not hover in the same position, and the drift will be observed in the vehicle.[lxxiii]

The above flight mode information will vary with the flight controller manufacturer. The ArduPilot-based flight controller systems will support mission planners as ground control stations, whereas boards based on other models and modules will have their own ground control stations.

The above-mentioned modes are quite basic and are important to know as these are the frequently used modes for application-based piloting.

Different flight controllers with different GCS have variations in settings, flight mode provided, etc. A ready-to-fly drone may have a predetermined flight mode whereas in a customised drone, with a need-based selection of an autopilot system, functions are adjustable and choices are more. The same applies to mission and flight planning. Not all methods of flying are provided in one drone. So, at times, it is possible that the drone may be used for a limited application if purchased without knowing the specific applications.

Note that there are a lot more settings available in the mission planner GCS. However, all the modes and settings are not explained here as we are learning the fundamental basics.

Chapter Insight

The ground control station is an application/software that allows you to connect your drone to a computer, do setups for your drone, plan missions, monitor your drone while flying and control it, get flight details and much more.

CHAPTER 11
Advantages of UAVs

11.1 Commercial Advantages

a) Drone-based Services

Landscape Analysis

The satellite remote sensing data can be used for the generation of Land Use/Land Cover (LULC) map of the region of interest of study which provides the moderate ground resolution of approximately 7.5m, whereas drones can capture images at very low altitude, collect micro-scale information and provide data with resolutions of 2cm to 10cm.[lxxiv] Precision flying with GPS/RTK captures data accurately such that it can minimise the errors to mm level.

Drones have the capability to save time and human efforts in certain commercial projects like land mapping, corridor mapping, thermal inspection, agriculture survey and sprayings, etc.

(Applications are explained in the next chapter.)

b) Decreased Risk and Dangers for Workers

Drones can easily be flown to certain tedious inspecting locations such as chimneys, refineries, pipeline inspections, transmission line inspections, solar farm inspections, wind turbines, etc. This eliminates the need for workers to carry out manual inspections by going into certain danger zones like chimneys and certain kilometres long pipelines.

c) Requires Less Time to Carry Out Missions

It requires less time to plan a mission, certain parameter settings and get ready for take-off.

d) Saves Time and Money

A ready-to-use drone and its accessories may not look affordable for purchase at times, but there are benefits in a way that a single drone can be used for different inspections, surveying and mapping applications as well as can capture images, high-resolution videos, etc.

e) Fast SOS Services

In the case of an emergency situation, a rapid task can be carried out in even remote areas. For example, during the need of an urgent organ donor, a donor or organ may not be available instantly in the same location. Under such conditions, the transfer of organs, blood or vaccines can be done using a drone.

11.2 Military Advantages

a) No Risk of Losing a Pilot

Drones can be handy, easy and capable of targeting enemy vehicles/locations during wartime. This could be called as a replacement for fighter aircraft. Maybe not in terms of speed, however, in terms of carrying out missions, surveillance and hitting out at the targets with bombs and guided-missile is what drones are capable of. Selfflying drones can save the life of a pilot when compared to an enemy hitting out at fighter aircraft.

b) **Low Risk of Defence Personnel**

A drone can patrol across the border autonomously without any requirement of defence personnel's presence at that location.

As Nano Drones are under research and even manufacturing, it is going to be useful for checking out on certain matters during terrorist's activity in danger zones.

c) **Disaster Management and Rescue Operations**

Drones can be used for monitoring at earthquake-affected sites, landslide areas in mountains and valleys, fire emergencies, during and post-flood for the analysis of the affected sites, even under critical situations.

Chapter Insight

Advantages and applications of a drone will look quite similar as they are interdependent. The innovative the application will be, the more advantageous the drone will be.

CHAPTER 12
Application of UAVs

12.1 Commercial Applications

a) Survey and Mapping

- Road widening
- Town planning and infrastructure development
- Area calculation
- Area elevation check for new transmission lines/bridges/corridor construction
- To count roadside houses, slums, towers, poles, trees, etc.
- Over the ground mining

All of the above listed and related activities are dependent on the RGB type of camera that captures images with a certain overlap. The image capturing process is known as "data acquisition." Once the data acquisition is done, the data is sent for data processing where the data is processed as per the project and the needs of the client.

b) Utility Inspections – Defect Identification

- Gas and oil pipeline leakage
- Refinery inspections

- Solar farm inspections
- Transmission tower inspections and defect identification
- Crack identification in chimneys

Some of the above-listed activities and related projects may need special and specific types of cameras like thermal cameras, which detect the temperature of the object. These types of cameras are mainly used for solar inspection and where temperature analysis is important.

c) 3D Modelling of Structures

Buildings, monuments, towers, etc. can be modelled in 3D with the drone data capturing at different angles and processing it. It requires certain techniques of flying to capture images or videos at different angles. Once the data is acquired, it needs to be processed for the generation of a 3D model.

d) Surveillance Purposes

Mounting a zoom camera allows a user to do long-term surveillance at a particular location and even in a planned repetitive circuit. This is useful for some industries as well as forest departments for anti-poaching surveillance. This application can be covered by an electro-optical infrared camera mounted on a high endurance drone.

e) Drone-based Deliveries

Deliveries of food, logistic parcels and first-aid are going to be operational soon. The use of drones for organs, blood, vaccines or small medical equipment transfers at high speed is also under process. This will resolve traffic or any other transfer related issues.

f) Stack Volume Analysis

In this application, the volume of coal and other material stacks can be calculated by carrying out 3D photography at multiple angles.

g) Agriculture and Spraying

Drones with multispectral cameras are capable of capturing images of the crops, which helps in analysing the plant health and the productivity of the soil. This also helps in measuring soil fertility, irrigation, crop yield, etc., and NDVI (Normalised Difference Vegetation Index) can be generated.

Drones with heavy payload capacities can be used for pesticide spraying in the farms. The pesticide sprayer drone can have a tank of 5litres/10 litres or even more, as per the design and capacity of the drone.

12.2 Military Applications

a) Border Security and Surveillance

Drones can be used for day and night patrolling across the borders with the help of certain different sensors. These drones are probably high endurance drones with a good range of operation.

b) Use in Warfare

Drones like Predators from the U.S. and Hermes from Israel are capable of targeting enemy vehicles on the ground as well as air with the help of bombs and guided missiles. So, these military drones and other upcoming advanced versions of them can replace manned fighter aircraft.

NOTE: By the time this book is being written, there is news from the UK Defence Secretary saying, "Only 10% of military aircraft will be manned by 2040." [lxxv]

Chapter Insight

Applications of the drones are far greater than what has been listed. The listed applications are of majority societal usage. The world needs more innovative ideas and applications from you.

Appendix A

Comparison of UAVs

A handy, "all-in-one" comparison of all types of drones to easily understand major applications, drawbacks and advantages.

Drone	Advantages	Disadvantages	Applications
Multirotor UAVs	• Can be plug-and-play in short time • Easy to operate • VTOL capability • Can be operated in a small area • Multiple applications	• Average endurance • Low payload capabilities	• Survey and mapping • Utility inspections • 3D modelling • Surveillance • Agriculture

(Contd.)

Appendix A

Drone	Advantages	Disadvantages	Applications
Fixed-Wing UAVs	• Higher payload capacity • Long endurance • High flying speed • Covers large areas	• Requires area for systematic take-off and landing • Cannot hover • Difficult to fly compared to Multirotor	• Survey and mapping Surveillance • Military purposes and defence
VTOL UAVs	• Higher payload capacity • Long endurance • High flying speed • Covers large areas • VTOL capabilities • Can hover	• Difficulties faced in transition period (due to improper configuration and wrong decision on altitude parameters)	• Delivery of vaccines, blood and other logistics
Helicopter UAVs	• Higher payload capacity • Long endurance • VTOL capabilities • Can hover	• Needs extraordinary flying skills	• Survey and mapping • Surveillance • Military purposes and defence • Agriculture spraying

Appendix B

Career Guidance

People are passionate about working on drones and flying them. It feels good to observe their enthusiasm while they are on the field or when they visit any such drone lab facilities. They feel that it is interesting to work on drones, and it really is. As **"Atmanirbhar Bharat"** campaign is a boost to Indian technologies and startups, I found Career Guidance necessary to be added to this book, as there is still confusion for the upcoming young engineers about where to start and what roles one can fit in. For them, here is brief information about for what roles require what jobs need to be done. This is very basic information and it may vary for different jobs in different companies.

a) **Research**

Drone and AI flying robots research is going ahead at a very fast pace where there can be a requirement from design to hardcore development stage as well as from programming to intelligently fly it in a weird environment. There are many breakthroughs which have not yet been achieved, and the world is waiting for skilled engineers and researchers that are multi-talented to perform certain experiments and invent something useful.

Aeronautical Engineers can go for research on design analysis, creating new efficient design, aerodynamics, propulsion systems, etc.

Mechanical Engineers can work on design and propulsion part as well as on analysis of design for their research. They can additionally go for research on materials.

Electronic Engineers are the best fit for research on electrical propulsion, flight control systems, communication systems, control-system parts of flight control and operations, simulations, payloads research, etc.

IT/Computer Engineers can have opportunities in flight controller systems, simulations, payload integration, operations research, AI-based research, etc.

b) **Manufacturing Management**

The manufacturing industry is growing, and they require technical managers to look after the manufacturing and assembly lines going well, plan the stages and process to produce efficiently. The technical manager in such position will inspect the parts defect, integration of the system level inspection as well as ground testing. For general management requirements, a qualified project management person can be the best fit. For technical inspection, an aeronautical as well as a mechanical engineer can be the best fit.

c) **Hardware Engineer**

Hardware engineer is for the design and development of the vehicle. The design part can provide the details of the electronics and accessories required to make it fly. All selections based on certain estimated calculations are included in the task of a hardware engineer. Additionally, this may include certain customisations on existing drones. In short, the drone gets ready to fly under a hardware engineer's supervision. An aeronautical engineer and a mechanical engineer can be the best fit.

d) Software Engineer

Computer/IT engineers and programmers can show their expertise in many ways with integration level coding for certain sensors and their operations, AI-based programming, increasing automatic flying level, output data format, integration of autopilot system with sensors, sensors integration with GPS, etc.

e) Design Engineer

In a manufacturing industry with unique design concepts, you need to have someone who can design drones on paper and 3D design. It is better if this person has the knowledge of aerodynamics. Design engineers can get the task of analysing the design on certain stress strain analysis platforms.

f) Repair and Maintenance

Wear-and-tear as well as crashes can happen on site. It is good that most of the time, our drones are in repairable conditions and to get it done, skilled technicians and junior engineers are required to perform multiple tasks related to replacement of parts.

g) Business Developer/Sales and Marketing Manager

For a company which manufactures drones or provides drone-based services, it is always important to sell the products.

This requires good marketing skills to be utilised by marketing or a sales manager to sell product and provide product support. They need to have good technical matter understanding capabilities, as technical products will have technical clients. This position is available in manufacturing organisations as well as survey companies.

h) **Project and Operation Management**

This is a management position, certainly not necessary to have a management professional working in this position. This position is for getting operations carried out smoothly and keeping in touch with operators for the work done, troubleshooting certain technical issues, managing projects based on priorities, managing spare requirements and inventory. This also includes flight planning, UAV maintenance observations, crash analysis, data quality control, etc. This job is good to be performed by an engineer experienced with certain aspects of operations, development, maintenance and managing team operations. This position is good to have for a drone-based survey company.

i) **Pilot/Operator**

Flying drones need attention and higher observation levels. The task includes manual as well as automatic flying skills with capabilities of mission planning, controlling the drone at different warning levels and emergency situations. The job requires good physical fitness, as hectic flying schedules are ahead.

This job can be done by any skilled pilot with any educational qualification level. However, it is necessary for any pilot to have a certain level of understanding with a classroom basic study carried out by instructors. This becomes helpful while troubleshooting on the field with the help of engineers from the head office.

Appendix C

Instructions for Technical Department

Care to be taken during Assembly and Integration

1. Drone parts are costly and need to be handled with care.
2. Balsa and carbon fibre materials are light in weight. So, they can break if proper care is not taken during assembly or fitment.
3. Utmost care should be taken while soldering battery connectors, power distribution boards, ESCs and any cable connectors.
4. Tightness of bolts on assembled parts should be inspected at regular intervals.
5. Care should be taken while connecting certain sensors with the autopilot system.
6. A proper method should be followed in case of changing the connector of a battery.
7. Battery should be placed properly on the drone.
8. Before starting calibration procedures, remove the propellers.
9. Care should be taken while connecting the battery and other connectors like battery connectors, flight controller connectors, motor connectors, etc.

10. Care should also be taken while disconnecting any connectors.

11. Proper distance should be maintained while carrying out the first test flight.

12. In case of any incident, disconnect the battery first.

Appendix D

Tools and Equipment to be Carried to the Field

Drone Parts, Tools and Equipment to be carried to the field for a Project/Mission

The products required for a project or mission depend on the type of mission, type of drone, sensors and other drone configurations. The list below can be used as part of the suggestion and the final list of materials required in the field can be prepared based on the tools and equipment missing as per your/user/company requirements.

Fixed-Wing (Engine/Electric Propulsion)

1. Drone fuselage
2. Wings (left and right)
3. Carbon rods (dowels)
4. Propellers: 1 no. + at least 1 spare propeller
5. LiPo batteries (as per the requirement)
6. RC transmitter and receiver pair
7. Remote control batteries
8. LiPo battery charger with connector cables

9. Laptop with charger

10. iPad/smartphone with charger

11. Paper tape

12. Double-sided tape

13. Insulation tape

14. Cable ties/zip ties

15. Adhesives: Epoxy and others

16. Twizzer

17. Hand drill

18. Sand paper

19. Cutter knife

20. Soldering set

21. Screw driver set

22. Allen Key set

23. Fuel mixture (if required)

24. Glow plug igniter (if required)

Multirotor Systems

1. Drone

2. Propellers: 4/6/8 no. + 1 full spare set propeller (depends on type of Multirotor)

3. LiPo battery

4. RC transmitter and receiver pair

5. Remote control batteries
6. LiPo battery charger with connector cables
7. Laptop with charger
8. iPad/smartphone with charger
9. Paper tape
10. Double-sided tape
11. Insulation tape
12. Cable/zip ties
13. Twizzer
14. Cutter knife
15. Soldering set
16. Screw driver set
17. Allen Key set

Appendix E

Instructions for Learners

The primary steps in the form of basic tips for flying are given below for Multirotor and Fixed-wing drones, separately. However, it is advised to take training from certified trainers or government approved trainers.

Multirotors

1. Have your fully charged battery ready.
2. Select an open area for practising flying where obstacles like trees and electric poles are less.
3. Find a well-levelled spotto take-off and land the drone.
4. The drone should be placed in such a way that the drone and your front match with each other.
5. Be careful while turning things on.
6. Turn on the remote controller first and then turn on the drone.
7. Keep a safe distance from the drone and take-off.
8. Your stick movement should be gentle and not reach extreme points of a stick for any moment. A slow stick movement will help you understand how much stick movement is to be applied in order to achieve a certain angle and position change.

Appendix E

9. Slowly practise up and down, right and left moments as well as moving forward and backward.

10. Practise yaw only when you are comfortable with other flying moments. During a 180° yaw, the drone faces you. At that time, the control of the drone gets reversed. For example, the right roll applied on the stick will move the drone towards your left.

11. Once you feel basic skills are learnt, try to make patterns like circle and square.

12. Land the drone on flat ground. Landing should be smooth. Do not catch the drones by hand.

13. Once landed, turn off the drone, and then turn off the remote controller.

Fixed-Wings

1. Have your fully charged battery ready.

2. Select an open area for practising flying where obstacles like trees and electric poles are less, probably the outskirts of the town.

3. Depending upon your model, you will need to determine the patch/runway area for taxiing take-off or hand launch of the drone.

4. The drone should be placed in such a way that the drone and your front match with each other.

5. Be careful while turning things on.

6. Turn on the remote controller first and then turn on the drone.

7. Check all control surface operations.

8. Keep a safe distance from the drone and take-off.

9. Your stick movement should be gentle and not reach extreme points of a stick for any moment. A slow stick movement will help you understand how much stick movement is to be applied in order to achieve a certain angle and position change.

10. Slowly practise right and left moments as well as pitch up moments.

11. Once you feel basic skills are learnt, try to make patterns like eight or a circle.

12. Land the drone on flat ground. Landing should be smooth.

13. Once landed, turn off the drone, and then turn off the remote controller.

Appendix F

Flight Logging Instructions and Checklist

A. Overall Instruction Set

1. A checklist of drones and accessories required in the field pertaining to the project should be prepared.

2. Beforehand, check if all the items listed are available with you.

3. Check if all the required batteries are charged and SD cards are formatted.

4. Listed items should be carried to the field.

5. Check site weather conditions.

6. Check whether the flying location is not falling under the no-flying zone or warning zone.

7. Pack the items carefully in such a way that they do not get damaged in transportation.

8. While flying, be attentive and continuously observe/monitor the drone on Ground control station and physically.

9. Prepare flight logs in hardcopy/soft copy such that they are legible enough and understandable. The flight log should include the name

of the location, project and date of the survey whenever a new flight log starts. An example of flight logging is given in Table 1 below.

| Project Name: |
| Location: |
| Flying Altitude: |
| Wind Condition: |
| Drone: |
| Date of Flying: |

Battery No.	Before Flight Voltage	After Flight Voltage	Start Time	End Time	Issue (if any)
01	98	55	13:00	13:15	Obstacle identified: drone returned to land; need to change flying altitude.

10. Check the site carefully before leaving. Do not miss anything on the field, in hotel rooms or in transportation vehicles.

11. After returning, check whether all the products assigned are available. In case anything is missing out, update the manager/director about the missing item.

Appendix F

12. Keep the critically low voltage battery in charging/storage, as per the requirement.

13. Note down the issues and changes required to be made.

B. At a Glance – Check List

Finalization of the Project

- ☐ Check weather forecast
- ☐ Drone Firmware updated
- ☐ Engine/Motor test run to the drone
- ☐ Remote controller charged
- ☐ Analyse flight area
- ☐ Check for permissions
- ☐ Ready fuel mixtures
- ☐ Laptop/mobile/iPad charged
- ☐ Prepare equipment list
- ☐ Remote controller firmware updated
- ☐ Check controls of drone
- ☐ Application updated
- ☐ SD card formatted
- ☐ Prepare flight plans
- ☐ Drone batteries charged
- ☐ All other accessories charged

Field Check before take-off

- ☐ Analyse area
- ☐ Unpack equipment and assemble
- ☐ Attach battery
- ☐ Turn on remote controller
- ☐ Calibrate compass
- ☐ Check controls
- ☐ Measure wind
- ☐ Check flight plan
- ☐ Select a take-off spot
- ☐ Remove gimbal lock
- ☐ Fill up the fuel
- ☐ Insert SD card
- ☐ Turn on drone
- ☐ Turn on the engine
- ☐ Check satellite strength
- ☐ Check return-to-home settings

During Mission

- ☐ Continuously monitor the drone
- ☐ Monitor battery percentage
- ☐ Land drone cautiously
- ☐ Keep checking signal strength
- ☐ Compare flight path with flight plan

Field Check after Landing

- ☐ Turn off the drone
- ☐ Remove SD card
- ☐ Empty the fuel tank
- ☐ Inspect damages
- ☐ Remove gimbal lock

- ☐ Turn off the remote controller
- ☐ Check data captured
- ☐ Remove battery
- ☐ Note down the damages
- ☐ Pack equipment

The above checklists are added here as a sample standard list. Points should be added or removed from the list as per the requirement.

Reference Links

i. Blackfoot Co., '1 Geronimo' paratroopers operate the RQ-11B Raven UAV at JBER by Arctic Warrior is licensed under CC BY 2.0

 https://www.flickr.com/photos/arcticwarrior/48639373406/in/photolist-2h76v9j-2h76vfb-2h74GWd-i3McQu-2h77e8A-GjUHgM-2h77eh8-i3Me93-2h77ekp-2h77emX-RD46L6-RD46tT-RD46q6-24MUqDa-RD46Fr-RD46nR-24MUqXB-RD46Ar-2fpsVFT-24MUqBX-2e1zhri-24MUqVx-24MUqLK-TgdukU-3qjwym-uFHyb5-rA9Qm9-fa7tvt-2hkgY2j-fvuowi-4EbepN-dUyaMr-Q6JXGA-tJB6Ja-5ECwvf-G7UExo-fuCh5U-5Eye9M-diaR9U-fmurvH-fmJCrA-diaQUQ-4kL4HE-fmJCmU-uDi8cC-fmJCwY-2eESDgG-8H6xaa-fmJCf1-5ASnG7

ii. https://www.jpl.nasa.gov/spaceimages/details.php?id=PIA19808

iii. 181212-N-PO203-0492 by Office of Naval Research is in public domain

 https://www.flickr.com/photos/usnavyresearch/32727782608/in/photolist-RS3rxW-RS3d39-f7AgJD-f7AfzT-2bepkey-f7QtMb-RS3fEo-RS3nhb-f7Qw3m-2cBfNsV-2cBfFR8-2bepsah-f7AiNM-JWhsjw-JWhruq-f7Airr-RS3wtN-KjryjU-Kjrxw1-JqNXKu-2e13BsT-2cU3Fsd-2hdnuHy-

Reference Links

2hdkaFp-2hdnHYL-JqNYqs-JqNVK7-9uh5xi-QeL4ci-vUPKve-RS3jc3-vRoiBf-puLN87-vPFKj9-N8WUcc-vSo7mH-f7AhDv-vzvvgB-vzoeN5-vzoeoN-JSg2CV-MFJruE-uV8pY8-J3KMZp-5Exyrn-vCddmS-J3H9oY-P1Khi3-JTet6S-Mbd3Ph

iv. "COMCAM_130622-N-OM642-389" by U.S. Naval Forces Central Command/U.S. Fifth Fleet is licensed under CC BY 2.0

https://search.creativecommons.org/photos/ba7e31fc-3aa0-42ae-aa24-37e5195dfda1

v. https://www.historytoday.com/archive/bombs-over-venice

vi. http://ffden-2.phys.uaf.edu/webproj/211_fall_2018/J-Rod_Maltos/history_3rd.html

vii. https://commons.wikimedia.org/wiki/File:Hewitt-Sperry_Automatic_Airplane_1918.jpg

viii. https://en.wikipedia.org/wiki/IAI_Scout

ix. https://en.wikipedia.org/wiki/General_Atomics_MQ-1_Predator

x. https://pixabay.com/photos/drone-flight-sky-flying-3723148/

xi. https://www.e-education.psu.edu/geog892/node/5

xii. https://dgca.gov.in/digigov-portal/jsp/dgca/homePage/viewPDF.jsp?page=InventoryList/headerblock/drones/D3X-X1.pdf

xiii. http://www.defenceimagery.mod.uk/fotoweb/fwbin/download.dll/45153802.jpg?ForceSaveDialog=no&D=A81D8793137680C0E66D08E8E606BD1F11732CAD1490F960D1DA871986B1F9663B985DAF642516CAAA2B3AA2B030D852712EB1B11DAC16910A2D0D9AC0

Reference Links

1 D B F D 4 F F 8 B 8 8 B B B 8 D F B 2 0 8 C A 3 E 8 D 3 8 A A 0 E D F 4 A 5 F 8 5 3 0 2 4 A 5 3 A 6 9 5 3AD0D33DD9360EBD10755885C649E1D77D3527A C2E94D2CBED32D1E4B20F3D1A4BB899F76D93 96B7199226538B04669FE449 2BD85E39C

xiv. https://www.flir.in/products/black-hornet-prs/

xv. https://www.independent.ie/world-news/and-finally/artificial-hummingbird-developed-26706343.html

xvi. https://commons.wikimedia.org/wiki/VTOL_UAV#/media/File:Nano_Hummingbird.jpg

xvii. https://pixabay.com/photos/drone-meadow-stands-flying-3525493/

xviii. https://www.questuav.com/drones/datahawk/#

xix. https://www.altiuas.com/transition/

xx. https://pixabay.com/photos/spraying-sugar-cane-sugar-cane-2746350/

xxi. "161009-N-PO203-371" by Office of Naval Research is licensed under CC BY 2.0 https://search.creativecommons.org/photos/6bfa9d08-0e8d-48ee-be3a-b1daac36b933

xxii. https://www.defenceimagery.mod.uk/fotoweb/Preview.fwx?position=1&archiveType=ImageFolder&albumId=5044&sorting=ModifiedTimeAsc&search=predator&fileId=EFCC51FEE65DA414D18085DA188CAB4552 4FFC4F7A63A403C47E17A8BEF1E554B796D6EA4F D91784A04B36049843E1FB56B129047A099FD2448D5A A2FD3EBB84D49852E5EF22F9F1E9930FDF2671F900 BC3FE983A7F78833A525EAE1E10D1F285D2914D

Reference Links

17479E705A7A12BC871A6188DE07DEE
FDC8451023F1040E151CCF64E2391B12E566375A1188C
B87C4B444E3C0866387014B4949AF6E71B7B167F9628240D
3674482835D33

xxiii. https://en.wikipedia.org/wiki/General_Atomics_MQ-9_Reaper

xxiv. "Hellfire Missiles_3890" by hoyasmeg is licensed under CC BY 2.0 https://search.creativecommons.org/photos/14d5e971-b5d0-493f-9d17-cb29e0983064

xxv. http://www.pilotfriend.com/training/flight_training/fxd_wing/emp.htm

xxvi. https://grabcad.com/library/x8-skywalker-uav-1

xxvii. https://store-en.tmotor.com/

xxviii. https://www.rc-airplane-world.com/glow-plugs-for-rc.html#:~:text=A%20glow%20engine%20(also%20often,methanol%2C%20oil%20and%20likely%20nitromethane.

xxix. https://encrypted-tbn0.gstatic.com/images?q=tbn%3AANd9GcQCbmbwNQUns1-cUvha5Vw3HVKez1iepvAsVg&usqp=CAU

xxx. https://en.wikipedia.org/wiki/Tricycle_landing_gear#/media/File:Cessna150taildraggerC-GOCB02.jpg

xxxi. www.suasnews.com

xxxii. PUMA Training by The USA Army is licensed under CC BY 2.0 https://www.flickr.com/photos/35703177@N00/8188591460

xxxiii. https://olive-drab.com/idphoto/id_photos_uav_rq2.php

xxxiv. DOI: 10.1109/ICUAS.2017.7991420

Reference Links

xxxv. http://dronecode.diyrobocars.com/wp-content/uploads/2017/06/falcon_vertigo_complete.jpg

xxxvi. https://commons.m.wikimedia.org/wiki/File:Ling-Temco-Vought_XC-142A.jpg#mw-jump-to-license

xxxvii. https://en.wikipedia.org/wiki/File:VMM-162_Osprey_on_the_tarmac_in_Iraq_on_April_1-2008.JPG (Public Domain)

xxxviii. 161012-N-JE250-047 by Office of Naval Research is licensed under CC BY 2.0 https://www.flickr.com/photos/usnavyresearch/ 29669467064/

xxxix. By ZullyC3P - Own work, CC BY-SA 4.0, https://commons.wikimedia.org/w/index.php?curid=49770593

xl. https://www.thanksbuyer.com/reptile-arrow-y3-tricopter-carbon-fiber-three-axis-multicopter-frame-for-gopro-fpv-26939

xli. https://www.google.com/url?sa=i&url=http%3A%2F%2Fwww.regimage.org%2Ftricopter-drone-kit%2F&psig=AOvVaw1_xxOqr_A_Ck6eLNZ4PNDM&ust=1612800369940000&source=images&cd=vfe&ved=0CA0QjhxqFwoTCIjbn_KT2O4CFQAAAAAdAAAAABAh

xlii. https://www.hindawi.com/journals/aaa/2014/320526/

xliii. https://www.alibaba.com/product-detail/JMRRC-Umbrella-Foldable-8-axis-carbon_60087393520.html

xliv. https://www.avlgear.com/xfold-rigs-spy-8urtf-octocopter-drone-rig-with-3-axis-gimbal

xlv. https://www.researchgate.net/publication/331298873_Autopilot_Design_for_a_Quadcopter/references

Reference Links

xlvi. https://hobbyking.com/en_us/s500-glass-fiber-quadcopter-frame-480mm-integrated-pcb-version.html?___store=en_us

xlvii. https://top-10-drones.com/build-first-drone-scratch-complete-guide-noobs/wood-frame-for-drone/

xlviii. www.rcpowerhour.info

xlix. https://control.com/technical-articles/a-technical-overview-of-drones-and-their-autonomous-applications/

l. https://www.mdpi.com/1424-8220/20/7/1940/htm

li. https://i.ytimg.com/vi/LXURLvga8bQ/maxresdefault.jpg

lii. https://circuitdigest.com/article/servo-motor-basics#:~:text=A%20servo%20motor%20is%20an,which%20run%20through%20servo%20mechanism.

liii. https://4.bp.blogspot.com/-HxMizyoQjfw/WPJWTyKilWI/AAAAAAAABK4/QnraFRQan1YWX_ILa8Pd1UMYFVpEvHdHwCEw/s1600/snap4arduino%2Bservo%2Bfuncionamiento.png

liv. https://www.dronetrest.com/t/power-distribution-boards-how-to-choose-the-right-one/1259

lv. 30A BLDC ESC Product Manual

lvi. https://www.robotbanao.com/wp-content/uploads/2020/03/71EJkrpCCyL._SL1500_-930x930.jpg

lvii. https://en.wikipedia.org/wiki/Lithium_polymer_battery

lviii. https://blog.ravpower.com/2017/06/lithium-ion-vs-lithium-polymer-batteries/

lix. https://www.epectec.com/batteries/lithium-vs-nimh-battery-packs.html

Reference Links

lx. https://www.batteryspace.com/LiFePO4/LiFeMnPO4-Batteries.aspx

lxi. www.hitecrcd.com

lxii. https://hitecrcd.com/products/chargers/discontinued-chargers-charging-accessories/x4-ac-plus-4-port-acdc-multi-charger/product

lxiii. https://www.flir.com/products/duo-pro-r/?model=436-0345-64-00

lxiv. https://micasense.com/rededge-mx

lxv. https://www.flysky-cn.com/products

lxvi. https://www.amazon.com/Radiolink-Transmitter-Controller-Multicopters-Helicopter/dp/B07FPF2HQR

lxvii. https://techterms.com/definition/gps

lxviii. https://www.apgsensors.com/sites/default/files/datasheets/IRU-3430.pdf

lxix. https://buy.garmin.com/en-US/US/p/557294

lxx. https://ardupilot.org/copter/docs/stabilize-mode.html#stabilize-mode

lxxi. https://ardupilot.org/copter/docs/altholdmode.html#altholdmode

lxxii. https://ardupilot.org/copter/docs/rtl-mode.html#rtl-mode

lxxiii. https://dl.djicdn.com/downloads/phantom_4/en/Phantom_4_User_Manual_en_v1.0.pdf

lxxiv. Kotaro Iizuka, Masayuki Itoh, Satomi Shiodera, Takashi Matsubara, Mark Dohar& Kazuo Watanabe | AnshumanBhardwaj

(Reviewing editor) (2018) Advantages of unmanned aerial vehicle (UAV) photogrammetry for landscape analysis compared with satellite data: A case study of postmining sites in Indonesia, Cogent Geoscience, 4:1, DOI: 10.1080/23312041.2018.1498180

lxxv. https://www.telegraph.co.uk/politics/2020/07/20/10-per-cent-military-aircraft-will-manned-2040-says-defence/